PREPARING THE WORKFORCE FOR DIGITAL CURATION

COMMITTEE ON FUTURE CAREER OPPORTUNITIES AND EDUCATIONAL REQUIREMENTS FOR DIGITAL CURATION

BOARD ON RESEARCH DATA AND INFORMATION

Policy and Global Affairs

NATIONAL RESEARCH COUNCIL
OF THE NATIONAL ACADEMIES

THE NATIONAL ACADEMIES PRESS
Washington, D.C.
www.nap.edu

THE NATIONAL ACADEMIES PRESS   500 Fifth Street, NW   Washington, DC 20001

NOTICE: The project that is the subject of this report was approved by the Governing Board of the National Research Council, whose members are drawn from the councils of the National Academy of Sciences, the National Academy of Engineering, and the Institute of Medicine. The members of the committee responsible for the report were chosen for their special competences and with regard for appropriate balance.

This study was supported by: Contract/Grant No. SLON 10000814 from the Alfred P. Sloan Foundation, Grant Number IMLS RE-04-11-0120-11 from the Institute of Museum and Library Services, and Grant Number NFS:OCI-1144157 from the National Science Foundation.
Any opinions, findings, conclusions, or recommendations expressed in this publication are those of the authors and do not necessarily reflect the views of the organizations or agencies that provided support for the project.

International Standard Book Number 13-978-0-309-29694-6
International Standard Book Number 10-0-309-29694-3

Additional copies of this report are available from the National Academies Press, 500 Fifth Street, NW, Keck 360, Washington, DC 20001; (800) 624-6242 or (202) 334-3313; http://www.nap.edu.

Copyright 2015 by the National Academy of Sciences. All rights reserved.

Printed in the United States of America

THE NATIONAL ACADEMIES
Advisers to the Nation on Science, Engineering, and Medicine

The **National Academy of Sciences** is a private, nonprofit, self-perpetuating society of distinguished scholars engaged in scientific and engineering research, dedicated to the furtherance of science and technology and to their use for the general welfare. Upon the authority of the charter granted to it by the Congress in 1863, the Academy has a mandate that requires it to advise the federal government on scientific and technical matters. Dr. Ralph J. Cicerone is president of the National Academy of Sciences.

The **National Academy of Engineering** was established in 1964, under the charter of the National Academy of Sciences, as a parallel organization of outstanding engineers. It is autonomous in its administration and in the selection of its members, sharing with the National Academy of Sciences the responsibility for advising the federal government. The National Academy of Engineering also sponsors engineering programs aimed at meeting national needs, encourages education and research, and recognizes the superior achievements of engineers. Dr. C. D. Mote, Jr., is president of the National Academy of Engineering.

The **Institute of Medicine** was established in 1970 by the National Academy of Sciences to secure the services of eminent members of appropriate professions in the examination of policy matters pertaining to the health of the public. The Institute acts under the responsibility given to the National Academy of Sciences by its congressional charter to be an adviser to the federal government and, upon its own initiative, to identify issues of medical care, research, and education. Dr. Victor J. Dzau is president of the Institute of Medicine.

The **National Research Council** was organized by the National Academy of Sciences in 1916 to associate the broad community of science and technology with the Academy's purposes of furthering knowledge and advising the federal government. Functioning in accordance with general policies determined by the Academy, the Council has become the principal operating agency of both the National Academy of Sciences and the National Academy of Engineering in providing services to the government, the public, and the scientific and engineering communities. The Council is administered jointly by both Academies and the Institute of Medicine. Dr. Ralph J. Cicerone and Dr. C. D. Mote, Jr., are chair and vice chair, respectively, of the National Research Council.

www.national-academies.org

COMMITTEE ON FUTURE CAREER OPPORTUNITIES AND EDUCATIONAL REQUIREMENTS FOR DIGITAL CURATION

MARGARET HEDSTROM, *Chair*, University of Michigan
LEE DIRKS, Microsoft Corporation (until August 2012) (deceased)
PETER FOX, Rensselaer Polytechnic Institute
MICHAEL GOODCHILD, University of California at Santa Barbara (retired)
HEATHER JOSEPH, Association of Research Libraries
RONALD LARSEN, University of Pittsburgh
CAROLE PALMER, University of Washington
STEVEN RUGGLES, University of Minnesota
DAVID SCHINDEL, Smithsonian Institution
STEPHEN WANDNER, The Urban Institute

Staff
SUBHASH KUVELKER, Study Director
PAUL F. UHLIR, Director, Board on Research Data and Information
DANIEL COHEN, Program Officer (on detail from the Library of Congress)
ALVAR MATTEI, Project Assistant
REBECCA HARRIS-PIERCE, Consultant

BOARD ON RESEARCH DATA AND INFORMATION

FRANCINE BERMAN, *Co-Chair*, Rensselaer Polytechnic Institute
CLIFFORD LYNCH, *Co-Chair*, Coalition for Networked Information
LAURA BARTOLO, Kent State University
HENRY BRADY, University of California at Berkeley
MARK BRENDER, Digital Globe Foundation
SAYEED CHOUDHURY, Johns Hopkins University
KEITH CLARKE, University of Southern California at Santa Barbara
KELVIN DROEGEMEIER, University of Oklahoma
CLIFFORD DUKE, Ecological Society of America
STEPHEN FRIEND, Sage Bionetworks
ELLIOT E. MAXWELL, e-Maxwell & Associates
ALEXA T. McCRAY, Harvard Medical School
ALAN TITLE, Lockheed Martin Advanced Technology Center

Staff
PAUL F. UHLIR, Director, Board on Research Data and Information
SUBHASH KUVELKER, Senior Program Officer
DANIEL COHEN, Program Officer (on detail from the Library of Congress)
ADRIANA COUREMBIS, Financial Associate
CHERYL WILLIAMS LEVEY, Senior Program Associate
ALVAR MATTEI, Program Assistant

PREFACE AND ACKNOWLEDGMENTS

In 2010 the Institute of Museum and Library Services (IMLS), a core sponsor of the Board on Research Data and Information (BRDI) of the National Research Council (NRC), had meetings with the members and the staff of BRDI which identified a need to examine the workforce issues for managing and enhancing the nation's digital assets in the coming decade. As a consequence of these discussions, the IMLS agreed to be the major sponsor of an NRC consensus study to address these concerns. The National Science Foundation (NSF) and the Alfred P. Sloan Foundation also agreed to sponsor the project.

The study titled "Future Career Opportunities and Educational Requirements for Digital Curation" was undertaken in fall 2011 by a study committee appointed by the president of the National Academy of Sciences. The committee was composed of experts in various fields, including library and information science, labor economics, domain sciences that rely heavily on digital data and information, higher education, and policy making from the government, academic, and private sectors. For the purposes of this study, "digital curation" is defined as "the active and ongoing management and enhancement of digital assets for current and future use."

The committee held four meetings that included open sessions for information gathering and input from all stakeholders. In conjunction with its second meeting, the committee organized a major national symposium—"Digital Curation in the Era of Big Data" in Washington, DC, on July 19, 2012. The symposium featured 10 invited speakers from the public and private sectors with expertise in digital curation in various fields. The results of that symposium were integrated into the study that led to this report.

This report has been reviewed in draft form by individuals chosen for their diverse perspectives and technical expertise, in accordance with procedures approved by the National Academies' Report Review Committee. The purpose of this independent review is to provide candid and critical comments that will assist the institution in making its published report as sound as possible and to ensure that the report meets institutional standards for objectivity, evidence, and responsiveness to the study charge. The review comments and draft manuscript remain confidential to protect the integrity of the process.

We wish to thank the following individuals for their review of this report: Suzanne Allard, University of Tennessee; Ruth Duerr, National Snow & Ice Data Center; Bryan Heidorn, University of Arizona; Charles Henry, Council on Library and Information Resources; Richard Luce, University of Oklahoma; Gary Marchionini, University of North Carolina; Maryann Martone, University of California at San Diego; Steven Miller, IBM Information Management Marketing; Jinfang Niu, University of South Florida; Charles Phelps, University of Rochester; Tomas Philipson, University of Chicago; and Gregory Withee, National Oceanic and Atmospheric Administration.

Although the reviewers listed above have provided many constructive comments and suggestions, they were not asked to endorse the conclusions or recommendations, nor did they see the final draft of the report before its release. The review of this report was overseen by Alexa McCray, Harvard University, and Carl Lineberger, University of Colorado. Appointed by the National Academies, they were responsible for making certain that an independent examination of this report was carried out in accordance with institutional procedures and that all review comments were carefully considered. Responsibility for the final content of this report rests entirely with the authoring committee and the institution.

The committee wishes to thank the following speakers at the symposium: Alan Blatecky, National Science Foundation; Vicki Ferrini, Columbia University; Joshua Greenberg, Alfred P. Sloan Foundation; Margarita Gregg, National Oceanographic and Atmospheric Administration; Myron P. Gutmann, National Science Foundation; Susan Hildreth, Institute of Museum and Library Services; Lawrence Hunter, University of Colorado at Denver; Anne Kenney, Cornell University; Elizabeth Liddy, Syracuse University; Andrew Maltz, Academy of Motion Picture Arts and Sciences; Nancy McGovern, Massachusetts Institute of Technology; Steven Miller, IBM; Michael Rappa, North Carolina State University; Michael Stebbins, White House Office of Science and Technology Policy; and David Weinberger, Senior Researcher, Harvard University. The committee also wishes to thank the following speakers at the open sessions of its meetings: Michael Chui, McKinsey Global Institute; Lauren Csorny, Bureau of Labor Statistics, U.S. Department of Labor; Rachel Frick, Council on Library and Information; Joshua Greenberg, Alfred P. Sloan Foundation; Gail Greenfield, National Research Council; Nirmala Kannankutty, National Science Foundation; Mimi McClure, National Science Foundation; Dane Skow, National Science Foundation; and Charles Thomas, Institute of Museum and Library Services.

The committee wishes to thank the IMLS, NSF, and Alfred P. Sloan Foundation for their sponsorship of this study. The committee also thanks Rebecca Harris-Pierce, who provided support as a consultant, and all members of the staff of the National Research Council who helped to organize the committee meetings and the symposium and to draft this report.

The committee dedicates this report to our colleague and friend Lee Dirks, who passed away in August 2012.

                                                             Margaret Hedstrom
                                                             Committee Chair

CONTENTS

Summary	1
Chapter 1 New Imperatives in Digital Curation and Its Workforce	7
1.1 The Digital Information Revolution	7
1.2 Dynamism in Use and Technology	8
1.3 Multiplicity and Diversity of Digital Information Handlers	9
1.4 Statement of Task	9
1.5 Definition of Terms	10
1.6 The Work and Workforce of Digital Curation: A Continuum	12
1.7 Scope of Topic and Time Frame	13
1.8 Organization of Report	14
1.9 References	15
Chapter 2 The Current State of Digital Curation	17
2.1 Evolution and Continuing Development of the Field of Digital Curation	17
2.2 Shared Norms and Standards	19
2.3 Best Practices for Good Quality	26
2.4 Curating for Durability	27
2.5 Further Advancement in the Field of Digital Curation	29

2.6 Impediments to the Advancement of the Field of Digital Curation	32
2.7 Measuring the Benefits of Digital Curation	33
2.8 Measuring the Costs of Digital Curation	35
2.9 Conclusions and Recommendations	38
2.10 References	41
Chapter 3 Current and Future Demand for the Digital Curation Workforce	**47**
3.1 Difficulties in Estimating Current Demand	47
3.2 Estimating Current Demand: Job Openings	49
3.3 Estimating Current Demand: Placements	53
3.4 Estimating Future Demand: Government Statistics	55
3.5 Automation and Future Demand	57
3.6 Conclusions and Recommendations	60
3.7 References	62
Chapter 4 Preparing and Sustaining a Workforce for Digital Curation	**63**
4.1 Envisioning the Education of Professional Digital Curators	63
4.2 Envisioning Education at the Other End of the Continuum	68
4.3 Current Educational Opportunities for Students of Digital Curation	69

4.4 Current Educational Opportunities for Student in Other Disciplines   71

4.5 Current Opportunities for Midcareer Employees   73

4.6 Looking Ahead   74

4.7 Building on Current Foundations   75

4.8 Many Stakeholders of Progress   76

4.9 Next Steps   76

4.10 Conclusions and Recommendations   78

4.11 References   80

APPENDIX A   83

APPENDIX B   87

SUMMARY

The massive increase in digital information in the last decade has created new requirements arising from a deficit in the institutional and technological structures and the human capital necessary to utilize and sustain the abundance of new digital information. This National Research Council consensus study report focuses on the need for education and training in digital curation to meet the societal demands for access to and meaningful use of digital information, now and in the future. For the purposes of this study, digital curation is defined as: "The active management and enhancement of digital information assets for current and future use." This definition provided the committee with a shared understanding of the scope of digital curation. As discussed below, digital curation entails more than secure storage and preservation of digital information because curation may add value to digital information and increase its utility.

There is no single occupational category for digital curators and no precise mapping between the knowledge and skills needed for digital curation and existing professions, careers, or job titles. The scope of digital curation is broader than that of data curation because digital curation includes all types of digital information. Digital curation differs from traditional curation of physical objects and collections because of the dynamic nature of digital information, its dependence on hardware and software for processing and analysis, its fragility, and many other characteristics.

The committee addressed the following issues in the course of this study. It identified the various practices and spectrum of skill sets that digital curation comprises, looking in particular at human versus automated tasks, both now and in the foreseeable future. It examined the possible career path demands and options for professionals working in digital curation activities and analyzed the economic benefits and societal importance of digital curation for competitiveness, innovation, and scientific advancement. In particular, the committee identified and analyzed the evolving roles and models of digital curation functions in research organizations and their effects on employment opportunities and requirements. It also identified and assessed the existing and future models for education and training in digital curation skill sets and career paths in various domains.

This report is organized into four chapters to address the Statement of Task, which is presented in full in Chapter 1, Section 1.4. Chapter 1 defines digital curation, establishes the scope of digital curation activities, and identifies factors that may influence workforce demand and the character of digital curation over the course of the next decade. Chapter 2 analyzes the current state of digital curation in detail, including the opportunities and benefits of digital curation and the character of digital curation work. Current and future demand for people with digital curation knowledge and skills along a spectrum from full-time professional curators to skills that anyone doing data-intensive work will need are examined in Chapter 3. Finally, Chapter 4 analyzes the current state of educational opportunities in digital curation and proposes measures for education and training to address the growing and unmet needs. The remainder of the summary is organized according to the main conclusions and recommendations of Chapters 2, 3, and 4.

Chapter 2 has five principal conclusions:

Conclusion 2.1: Demands for readily accessible, accurate, useful, and usable digital information from researchers, information-intensive industries, and consumers have exposed limitations, vulnerabilities, and missed opportunities for science, business, and government, as a result of the immaturity and ad hoc nature of digital curation. There is also a push for greater openness and transparency across many sectors of society. Taken together, these factors are creating an urgent need for policies, services, technologies, and expertise in digital curation. Although the benefits of digital curation are poorly understood and not well articulated, significant opportunities exist to embed digital curation deeply into an organization's practices to reduce costs and increase benefits.

Conclusion 2.2: There are many inducements that could drive advances in the field of digital curation:
- Organizations that can serve as leaders, models, and sources of good curation practices;
- Government requirements for managing, sharing, and archiving information in digital form;
- Protection of digital assets to build trust and satisfy consumers and to maintain competitiveness in business and scientific research;
- Rewards and professional recognition for the value that curation adds to digital information; and
- Pressure from consumers, citizens, and society at large for accountability and transparency in business and government.

Conclusion 2.3: There are also barriers to developing the capacity for comprehensive, affordable, and effective digital curation. Some impediments, such as attitudes about sharing data and concerns over privacy, competitive advantage, security, and misuse of digital information, are difficult to delineate or measure. Insufficient financial resources for digital curation are a commonplace concern.

Conclusion 2.4: Cost models and studies of digital curation costs consistently identify human resources as the most costly component of digital curation. Current cost models are likely to underestimate the costs of curation tasks performed by the creators and producers of digital information, because no techniques have been developed to segregate or measure curation costs prior to accessioning into a repository. There is a pressing need to identify, segregate, and measure the costs of curation tasks that are embedded in scientific research and common business processes.

Conclusion 2.5: Although standards and good practices for digital curation are emerging, there is great variability in the extent to which standards and effective practices are being adopted within scientific disciplines, commercial enterprises, and government agencies. The absence of coordination across different sectors of the economy and different organizations has led to limited adoption of consistent standards for digital curation and resulted in the fragmented dissemination of good practices.

Four recommendations flow from these conclusions:

Recommendation 2.1: Organizations across multiple sectors of the economy should create inducements for and lower barriers to digital curation. The Office of Science and Technology Policy (OSTP) should lead policy development and prioritize strategic resource investments for digital curation. Leaders in information-intensive industries should advocate for the benefits of digital curation for product innovation, competitiveness, reputation management, and consumer satisfaction. Leaders of scientific organizations and professional societies should promote mechanisms for recognition and rewards for scientific and professional contributions to digital curation.

Recommendation 2.2: Research communities, government agencies, commercial firms, and educational institutions should work together to accelerate the development and adoption of digital curation standards and good practices. This includes (1) the development and promotion of standards for meaningful exchange of digital information across disciplinary and organizational boundaries; and (2) interoperability between systems used to collect, accumulate, and analyze digital information and the repositories, data centers, cloud services, and other providers with long-term stewardship and dissemination responsibilities.

Recommendation 2.3: Researchers in economics, business analysis, process design, workflow, and curation should collaborate to identify, estimate or measure, and predict costs associated with digital curation. The National Science Foundation, the Institute of Museum and Library Services, relevant foundations, and industry groups should solicit proposals for and fund such research.

Recommendation 2.4: Scientific and professional organizations, advocacy groups, and private-sector entities should articulate, explain, and measure the benefits derived from digital curation, including "after-market" benefits, risk mitigation, and opportunities for private-sector investment, innovation, and development of curation technologies and services. The benefits should include outcomes that generate measurable value as well as less tangible benefits such as the accessibility of digital information over time for scientific research, organizational learning, long-term trend analysis, policy impact analysis, and even personal entertainment. Such research is necessary for the development and testing of sophisticated cost-benefit (or cost-value) models and metrics that encompass the full range of digital curation activities in many types of organizations.

Chapter 3, which looks at the demand side of the equation, presents five main conclusions:

Conclusion 3.1: Jobs involving digital curation exist along a continuum, from those for which almost all tasks focus on digital curation to those for which digital curation tasks arise occasionally in a job that is embedded in some other domain.

Conclusion 3.2: Although digital curation is not currently recognized by the Bureau of Labor Statistics in its Standard Occupational Classification, other sources of employment data identify the emergence and rapid rise of digital curation and associated job skills.

Conclusion 3.3: There is a paucity of data on the production of trained digital curation professionals and their career paths. Tracking employment openings, enrollments in professional education programs, and the placement and career trajectories of graduates from these programs would help balance supply with demand on a national scale.

Conclusion 3.4: The pace of automation and its potential impact on both the number and types of positions that require digital curation knowledge and skills is a great unknown. Automation of at least some digital curation tasks is desirable from a number of perspectives, and its potential has been demonstrated in several domains.

Conclusion 3.5: Enhanced educational opportunities and new curricula in digital curation can help to meet the rapidly growing demand. These opportunities can be developed at all levels and delivered through formal and informal educational processes. Digital learning materials that are accessible online, for example, may achieve broad exposure and possible rapid adoption of digital curation procedures.

We make two recommendations in light of the analysis in Chapter 3:

Recommendation 3.1: Government agencies, private employers, and professional associations should develop better mechanisms to track the demand for individuals in jobs where digital curation is the primary focus. The Bureau of Labor Statistics should add a digital curation occupational title to the Standard Occupational Classification when it revises the SOC system in 2018. Recognition of digital curation as an occupational category would also help to strengthen the attention given to digital curation in workforce preparation.

Recommendation 3.2: Government agencies, private employers, and professional associations should also undertake a concerted effort to monitor the demand for digital curation knowledge and skills in positions that are primarily focused on other activities but include some curation tasks. The Office of Personnel Management should issue guidelines for specifying digital curation knowledge and skills that should be included in federal government position descriptions and job announcements. Private employers, professional associations, and scientific organizations should specify the digital curation knowledge and skills needed in positions that require them.

The final chapter offers three broad conclusions:

Conclusion 4.1: Although the number and breadth of educational opportunities supporting digital curation have grown, existing capacity is low, especially for the initial education of professional digital curators and the midcareer training of professionals with credentials in another field. In particular:

• Graduate and postgraduate certificate programs for educating professional digital curators (e.g., in Library and Information Schools and iSchools) are expanding, but workforce demand is projected to exceed the output of existing programs.

• Midcareer practitioners with little or no formal education in digital curation rely on a spectrum of types of training, including online and in-person, experimental and time-tested, and just-in-time training, but this too is not sufficiently developed.

Conclusion 4.2: The knowledge and skills required of those engaged in digital curation are dynamic and highly interdisciplinary. They include an integrated understanding of computing and information science, librarianship, archival practice, and the disciplines and domains generating and using data. Additional knowledge and skills for effective digital curation are emerging in response to data-driven scholarship. More specifically:

• Individuals with an undergraduate degree in science, technology, engineering, or mathematics (STEM) disciplines and graduate-level education in digital curation are—and will continue to be—in particular demand as digital curators.

• Discipline specialists with informatics and digital curation expertise are, and will continue to be, in demand to provide discipline-focused curation services.

• Although the multidisciplinary character of digital curation as a career currently suggests a graduate education level, some knowledge and skills may be acquired through 2-year associate or 4-year bachelor's degrees.

• Continuing professional education alternatives will need to be flexible and diverse, providing a range of introductory and more specialized options through several modes of delivery, such as workshops, tutorials, online course modules, and webinars.

Conclusion 4.3: The range of needs and opportunities in digital curation, particularly when reflected in Office of Personnel Management position descriptions and Bureau of Labor Statistics descriptions of occupations, will require building and advancing a diverse community supported by a core of professionals and practitioners.

In light of these conclusions, we make the following three recommendations:

Recommendation 4.1: OSTP should convene relevant federal organizations, professional associations, and private foundations to encourage the development of model curricula, training programs and instructional materials, and career paths that advance digital curation as a recognized academic and professional discipline.

Recommendation 4.2: Educators in institutions offering professional education in digital curation should create cross-domain partnerships with educators, scholars, and practitioners in data-intensive disciplines and established data centers. The goals of these partnerships would be to accelerate the definition of best practices and guiding principles as they evolve and mature, to help ensure that educational and training opportunities meet the needs of scientists in specific disciplines, analysts in different business sectors, and members of other communities utilizing digital curation systems and services.

Recommendation 4.3: Federal agencies, private foundations, and industrial research organizations should foster research on digital curation that makes fundamental progress on problems with practical applications in their respective domains. Initial activities should focus on establishing research priorities and baseline analyses, including engagement and outreach through:

- Conferences and symposia designed to recognize and communicate the need for, benefits of, and successes in digital curation; and
- Workshops for researchers in the public and private sectors to develop coordinated research agendas focused on enhancing the value and utility of digital resources, including metadata, interoperability, and automation.

The resulting agendas for research in digital curation should be tightly coupled with the curricula and offerings of educational programs to shape the field during a time of dynamic and dramatic growth and change.

Chapter 1

New Imperatives in Digital Curation and Its Workforce

1.1 The Digital Information Revolution

A revolution in digital information is occurring across all realms of human endeavor. This revolution is characterized by a tremendous increase in the quantity of digital information, dynamism in both the purposes for and technology by which digital information is used, and multiplicity of digital information handlers. The implications of this revolution in digital information include an imperative for effective long-term digital curation and a workforce sufficient in skill and number to meet that challenge.

The vast increase in the quantity of digital information is made possible by computer and networking technologies that create, capture, copy, share, and store massive amounts of information easily and at very low cost. Many studies examining the production, use, and sharing of digital information confirm an astounding rate of increase (Lyman and Varian, 2003; Bohn and Short, 2009; IDC, 2014). The increase is occurring across all sectors, from scientific research to government administration, health care, business, and cultural and personal expression.

Research in the sciences is producing an enormous and rapidly growing flow of digital information. From distant satellites to medical implants, a great range of sensors permits the collection of unprecedented quantities of digital information across the scientific disciplines. That vast quantity in turn creates challenges of how best to share, store, manage, and analyze the data. Indeed, the availability of big data has transformed many scientific disciplines, with fields such as molecular biology, biodiversity, ecosystem studies, and geography becoming very data intensive. Hybrid fields such as bioinformatics, biodiversity informatics, ecoinformatics, and geospatial science have also emerged in response to the immense quantities of digital information.

Government is another source of the huge increase in digital information. The federal government generates a tremendous flow of administrative documents from agencies such as the Social Security Administration, Medicare, Medicaid, the Veteran's Administration, and the Internal Revenue Service, while state and local governments produce records in such areas as education, voter registration, and property ownership. Federal and state governments also generate a large volume of statistical data that are collected and disseminated for policy research (Card, D. R. et al., 2010).[1]

---

[1] The major federal statistical agencies include the Census Bureau, Bureau of Labor Statistics, Bureau of Transportation Statistics, Bureau of Justice Statistics, National Center for Education Statistics, National Agricultural Statistics Service, and the National Center for Health Statistics. In addition, the Office of Management and Budget identifies approximately 80 other federal agencies and organizations that produce statistics (Office of Management and Budget, 2012).

The health sector also contributes to the unwieldy quantity of digital information. With the increased use of digital imaging technology and the shift to electronic health records, health care providers and insurance companies are recording detailed information on patients, diagnoses, treatments, and payments. The increased flow of digital information is used to analyze the efficacy of treatments, trends in diseases and chronic conditions over time, and costs of various procedures and treatments.

Collection of vast amounts of digital information is ubiquitous in the private sector as well. Retailers, banks, credit-rating agencies, and insurance companies record transactions as digital information. In the entertainment industry, digital media are the primary, and in some cases, the exclusive mode for distributing products, whether texts, music, games, or motion pictures (Academy of Motion Pictures Arts and Sciences, 2007). Commercial strategies have been transformed by the use of digital information. Many companies use data analytics, web analytics, and other business intelligence techniques to analyze consumer behavior, target advertising toward specific individuals or consumer groups, manage inventory and production schedules, and devise business strategy (Manyika et al., 2011).

Private citizens, too, are creating and sharing enormous amounts of digital information. Social media are platforms for huge quantities of photos, videos, and personal information, much of it ephemera, yet also perhaps of cultural and historical value or useful in social research (Lee, C. A., ed. 2011). According to an estimate of information generation in 2012, every minute of the day users uploaded 48 hours of new video to YouTube, sent 204,166,667 e-mail messages, and submitted over 2 million search queries to Google (Spencer, 2012).

Much of the increased flow of information is "born digital." Research data from sensors as varied as particle accelerators, astronomical observatories, remote sensing platforms, or automated DNA sequencers are captured in digital formats. Government, commercial, and health records are initiated in digital formats. A variety of technologies permit personal and cultural creative expression as digital information. A substantial portion of this vast new trove of digital information also results from major initiatives to digitize analog data, from historic maps and weather almanacs to audio recordings, photographs, and even the label data on museum specimens. Libraries, archives, and museums are transitioning from physical to digital collections and from manual to automated processes for collections management.

One further component of the vast increase in digital information is metadata. Metadata, or data about data, describe the contexts and the content of data files. Accurate and complete metadata are essential for analyzing data and can themselves be an important resource for research. In some instances, the volume of metadata required for effective documentation exceeds the volume of the data being described.

This vast increase in the sheer quantity of digital information—from whatever source or sector, whether originating in or transferred to digital format, and whether consisting of data or metadata—presents many challenges for digital curation. The capture, management, preservation, and storage of content this large require significant curatorial skills and knowledge.

1.2 Dynamism in Use and Technology

In addition to the massive increase in the sheer quantity of digital information, other characteristics of the revolution in digital information demand curatorial expertise. Much digital information is being used or reused in ways not anticipated when that information was collected. Digital information moves readily across temporal boundaries, as digitized data from ships' logs in the seventeenth century are analyzed by climate scientists in the twenty-first. Digital information also moves across sectorial boundaries, as epidemiologists examine commercial data on consumer searches for flu remedies. Digital information ignores disciplinary boundaries as well, as researchers in bioinformatics combine datasets originating in biology, genetics, and engineering. Such dynamism in the use and reuse of digital information places high demand on curatorial expertise. The curatorial challenges of interoperability and accessibility are great when the uses of digital information are so dispersed and fluid.

The technology for capturing, managing, and storing digital information is also in continual flux. Both the hardware and software for accessing, interpreting, and preserving digital information are continually being upgraded. Curatorial strategies for storage and retrieval are therefore never definitively settled. Dependencies between data, software, and metadata also raise challenges for curation. In a rapidly shifting technological environment, software and metadata can change independently of the data; old software quickly becomes unusable and old metadata become impossible to interpret. Anticipating such problems and developing strategies to mitigate them are a core activity of digital curation.

1.3 Multiplicity and Diversity of Digital Information Handlers

The revolution in digital information may also be characterized by the multiplicity and diversity of people handling digital information and the contexts in which they do so. Handlers of digital information include professionals trained and engaged in curation per se, as well as experts in a variety of domains who must do some digital curation in order to accomplish other aims. But there is also an enormous range of others—librarians, administrators, and those participating in crowdsourcing—who are producing or gaining access to stores of digital information. They do so in a variety of organizational contexts, from enormous government-funded data repositories to smaller libraries, from major commercial databanks and cloud services to startup companies, as well as in private archives and collections.

The multiplicity and diversity of people involved with digital information and the contexts of that involvement have implications for digital curation and curators. Different levels of curation might be appropriate to different types of producers and users of digital information. Not only the technology of access to digital information but also the propriety of it is a concern, with curators having to address issues of data security. Methods and approaches of digital curation will also need to be adjusted in response to the tremendous range of resources, methodologies, and organization of workflows in the very different settings where digital curation activities occur.

1.4 Statement of Task

The revolution in digital information requires an accompanying surge in the advancement of digital curation, and therefore in the digital curation workforce. How to meet the demand for a digital curation workforce, suitable both in expertise and in number, is the challenge the study committee addressed. The specific charge to the study committee, as described in the Statement of Task, is as follows:

1. Identify the various practices and spectrum of skill sets that digital curation comprises , looking in particular at human versus automated tasks, both now and in the foreseeable future.
2. Examine the possible career path demands and options for professionals working in

digital curation activities, and analyze the economic and social importance of these employment opportunities for the nation over time. In particular, identify and analyze the evolving roles and models of digital curation functions in research organizations, and their effects on employment opportunities and requirements.
3. Identify and assess the existing and future models for education and training in digital curation skill sets and career paths in various domains.
4. Produce a consensus report with findings and recommendations, taking into consideration the various stakeholder groups in the digital curation community, that address items 1–3, above.

The remainder of this chapter defines digital curation and its key elements, reflects further on the characteristics of digital curation and its workforce, delineates the topical scope and time frame of the committee's work, and presents the organization of this report.

1.5 Definition of Terms

1.5.1 Digital Curation

For the purposes of this study, digital curation is defined as *the active management and enhancement of digital information assets for current and future use.* After reviewing numerous alternatives, the committee adopted this definition so as to encompass a wide range of curatorial activities and practices. This section considers the term digital curation itself and elaborates on each element of the definition.

Digital curation differs in several ways from curation as it is traditionally understood. Generally, curation denotes the selection, care, and preservation of collections of objects. The content of curated collections is typically relatively small, consisting of rare or unique works of art, rare books and manuscripts, important natural and physical specimens, or cultural artifacts. Curation takes place in relatively limited organizational contexts: libraries, archives, museums, art galleries, herbaria, and similar institutions; the work of specially trained curators has focused primarily on preserving and archiving collections within these settings. Digital curation displays some continuity with this tradition of curation. Regardless of whether a collection is physical or digital, a curator must appraise its value and relevance to the community of potential users; determine the need for preservation; document provenance and authenticity; describe, register, and catalog its content; arrange for long-term storage and preservation; and provide a means for access and use.

Yet digital information also poses many new challenges for curation: the immense and ever-increasing quantities of material to be curated, the need for active and ongoing management in a context of continually changing uses and technology, and the great diversity of organizational contexts in which curation occurs.

1.5.2 Active Management and Enhancement

The phrase "active management and enhancement" was chosen to distinguish curation from simply collecting and storing data and information. Active management denotes planned, systematic, purposeful, and directed actions that make digital information fit for a purpose. It includes coordinated activities that allow users to understand and exploit digital information assets and to ensure their integrity over time. Active management also refers to activities that ensure that digital information will remain discoverable, accessible, and useable for as long as potential users have a need or a right to use it. It may further involve securing digital information from unauthorized access.

Active management of digital information entails a wide range of both managerial and technical activities. Relevant managerial activities include developing policies for digital curation; assessing risks to the organization that might result from current technology, policies, and curation practices; identifying information assets; evaluating the effectiveness of systems and processes that support digital curation; monitoring compliance with regulations and best practices; mobilizing financial and technical resources for curation; and recruiting and training qualified digital curation personnel to support consistent curation practices across an organization. Technical activities include working directly with the hardware and software systems that support information management, such as establishing and operating repositories for long-term archival management of digital information, organizing and cataloging digital information assets, creating or enhancing the metadata associated with digital information objects, disseminating digital information, and managing access to repositories and their content.

Enhancement means taking measures to increase the value of digital information for current and future use. Most digital information is neither naturally useful nor immediately valuable at the moment it is created or collected. Curation processes that reduce or eliminate noise in the data and that detect and correct errors or other anomalies may increase its immediate utility. Data may need to be repackaged to prevent format obsolescence or represented in a form that satisfies the needs of specific user communities. Enhancement does not, of course, include intentional manipulation to support false conclusions.

The collection or assembly of descriptive metadata is another very important aspect of enhancement. Transforming digital data into useful information usually requires active intervention by skilled people and software applications. Furthermore, because digital information is fragile, corruptible, easily altered, and subject to accidental and intentional deletion, maintaining the integrity of information is a critical aspect of digital curation. Digital curation can enhance the integrity of digital information and increase its trustworthiness through security and restricted access to curation systems, replication, documentation of any transformations of the information, and auditable process and procedures.

1.5.3 Digital Information Assets

Assets have value. Not all digital information is an asset. In a stream of ephemera and communication, determining which digital information constitutes an asset, as opposed to a liability, an intermediate product, or just plain noise, is highly dependent on the context in which it is used or is anticipated to be used. Further, some digital information can become an asset through curatorial activity—not only through enhancement of its utility, but by measures to ensure ease of discovery, access, and distribution.

1.5.4 Current and Future Use

That digital information has both current uses and potential future uses has important implications for digital curation. The range of current uses across many sectors requires curation of digital information for that contemporaneous diversity of users and methodologies. Future use of digital information, both within and beyond the context in which it was first created or collected, places additional demands on curation. Attention must be paid to updating and upgrading technologies, software, and metadata, both for the preservation of the digital information and for maintenance of access to it.

1.6 The Work and Workforce of Digital Curation: A Continuum

To further introduce the committee's approach to the topic of preparing a workforce for digital curation, some attributes of how, by whom, and where that work is conducted merit comment. An essential attribute of the work of digital curation is that it is accomplished along a continuum. It does not consist of a discrete profession labeled "digital curator" with a defined set of tasks undertaken in a dedicated setting. Rather, it is more usefully conceived as a series of activities undertaken by a range of personnel in a great variety of settings. This heterogeneity has major implications for measuring the work of digital curation, estimating demand for its workforce, and determining how best to train that workforce.

The continuum of professional positions including some responsibility for digital curation is very long. At one end of that continuum are specialists whose jobs consist exclusively or primarily of curation. They are designated personnel with specific expertise and training in the field of digital curation. At the opposite end of the continuum are jobs that may include curatorial tasks from time to time. Digital curation may be an essential but not predominant part of these jobs. The curatorial activities included in these jobs may be deemed a chore, even a distraction from the primary work to be accomplished; they may not even be recognized as curation.

Importantly, the two ends of that continuum are connected. Most digital information derives its meaning, value, and utility in relation to the domains, problems, or processes to which it is applied. Therefore, professional curators cannot make sound decisions, provide services, or add value to digital information without some knowledge of those domains, problems, or processes. In scientific fields, for example, digital curators need familiarity with the terminology, methods, common data types and formats, standards of acceptance, and norms of a specific scientific community. In commercial environments, digital curators need knowledge of the competitive environment, regulatory framework, and nomenclature of a particular line of business. At the other end of the continuum, professionals such as research scientists or marketing analysts who are engaged in work that is seen as far from curatorial must nonetheless be proficient in curation. Without the knowledge and skills to conduct sound curatorial practices, such as recording metadata or maintaining accessible formats or properly combining datasets, those professionals will fail at the rest of their work.

The organizational and institutional settings in which digital curation is accomplished also vary along a large spectrum. Some of these, such as formal data centers and repositories or government statistical agencies, may be explicitly dedicated to the curation of digital information. They may pursue curation as an end in itself. In other settings, the curation of digital information may be but one component of a very broad set of activities, in which curation serves but does not define the goal. There is variation all along the spectrum. In some organizational settings, the work of curation will be concentrated among specific personnel; in others it will be dispersed. Some settings will take responsibility for digital information from the moment it is collected or created, whereas others will begin to manage it only after the original producer has assigned metadata, or after the original user no longer has a need for it.

The various organizations and institutions in which digital curation is conducted also have very different resources, and therefore very different ways of organizing and accomplishing the work of curation. Investment in technology and human capital vary. The potential for automating some activities of digital curation is also variable. It may be affected by such factors as the size of organizations, the types of technical systems in place, the volume and types of information, and the degree to which curation tasks have been integrated into workflows and business processes. These variations can be found not only between different sectors (e.g., financial, retail, entertainment, manufacturing, health care, research, and education), but also within organizations in the same sector.

Digital curation is not the only term used to characterize activities that enhance the value of information. *Information management*, *data management*, *data stewardship*, *data governance*, and *digital archiving* are related terms used to describe processes and activities that overlap with curation. Information management is concerned with the full range of issues that affect acquisition, organization, processing, and delivery of information including efficiency of operations, controlling costs, and regulatory compliance, often in the context of an organization-wide information architecture. According to Data Management International (DAMA International[2]), data management is the "development and execution of architectures, policies, practices and procedures that properly manage the full data lifecycle needs of an enterprise." Information management and similar fields entail processes and activities that overlap with curation, yet are distinct from it. What distinguishes curation from these other fields is its emphasis on enhancing the value of information assets for current and future use and its attention to the repurposing and reuse of information, both within and beyond the context in which it was first created or collected.

In the absence of a formal occupational classification of digital curator responsible for a delimited set of tasks in a standard work setting, the committee's approach was to identify and analyze digital curation activities, investigating different scenarios for distributing digital curation activities. The way responsibility for digital curation activities is distributed across and within organizations will be an important factor in determining the right mix of curation knowledge and skills in the workforce. The committee's approach also reflects the dynamic nature of digital curation, in which standards and best practices are still evolving and automation lags behind the exponential growth in digital information.

1.7 Scope of Topic and Time Frame

To minimize confusion over the scope of activities and the necessary knowledge and skills that digital curation comprises, the committee reached a consensus about some fields that are related to digital curation, but beyond the scope of this report. Data science and data analytics are two related fields, which, like digital curation, are recent and lack clear definitions and boundaries. The committee understood data science to mean the application of mathematics, statistics, and computer science to extract meaning from data and solve complex problems using statistical techniques, algorithms, and visualization. Data analytics extends statistical analysis with descriptive and predictive models to obtain knowledge from data by using insight from analyses to identify trends, evaluate performance, characterize consumer behavior, detect anomalies, recommend action, or guide and communicate decision making.

After reviewing definitions of these new jobs, the committee decided that positions that focus exclusively or primarily on developing algorithms, refining statistical techniques, mining data, and applying analytics to data did not fall within the committee's definition of digital curation. Digital curation differs from data science and analytics because curation is needed for many types of digital information, such as websites, blogs, social media, music, videos, geospatial information, online publications, and textual databases, to name but a few examples.

---

[2] http://www.dama.org/.

Furthermore, data analytics and data science typically focus on the immediate use of data for scientific and commercial purposes, rather than on current and potential future use of digital information.

The statement of task asked the committee to identify and analyze the evolving roles and models of digital curation functions in research organizations. The report pays particular attention to digital curation capacity and needs in research environments where recent changes in policy may raise the visibility of digital curation.

A further aspect of the scope of this report follows from the committee's decision to address digital curation activities rather than limit itself to examining the narrower occupational category of digital curator. Although emerging career paths of professional digital curators were investigated, it was not possible to analyze the economic and social importance of employment opportunities occurring across the full range of digital curation activities. The committee recognized that digital information flows readily across national borders through interconnected global infrastructures. Nevertheless, the committee determined that in order to make its task tractable, the primary focus of the report would be on workforce and educational needs for digital curation in the United States, while drawing on evidence from other countries and international efforts for salient examples and purposes of comparison.

Regarding the time frame, the committee was asked to consider the spectrum of knowledge and skills needed for digital curation now and into the foreseeable future. Given the dynamic nature of information technology and uncertainty about the speed at which organizations will develop policies, systems, and good practices for digital curation, the committee defined its time horizon as the next decade. Even within a 10-year time frame, numerous unknowns may influence the nature of digital curation activities and the demand for individuals with digital curation knowledge and expertise.

1.8 Organization of Report

This chapter has characterized the revolution in digital information, defined digital curation, and reflected on some characteristics and contexts of curatorial work. It has also delineated the Statement of Task and clarified the scope of what the committee undertook in order to address that task. Chapter 2 examines the evolution, current state, and ongoing development of digital curation. It also considers how to measure the benefits and costs of digital curation. Chapter 3 uses a variety of resources to devise estimates of current and future demand for the workforce in digital curation. Chapter 4 addresses the education of a workforce sufficient to meet the varied challenges of digital curation as they arise across different sectors and domains, within different organizational settings, at many different levels. It identifies and assesses the current state of educational opportunities in digital curation and considers steps for future progress.

1.9 References

Academy of Motion Picture Arts and Sciences, Science and Technology Council. 2007. *The Digital Dilemma: Strategic Issues in Archiving and Accessing Digital Motion Picture Materials.* http://www.oscars.org/science-technology/council/projects/digitaldilemma/.

Bohn, R. E., and J. E. Short. 2009. *How Much Information? 2009 Report on American Consumers*. December. Global Information Industry Center, University of California at San Diego. http://hmi.ucsd.edu/pdf/HMI_2009_ConsumerReport_Dec9_2009.pdf. Accessed June 16, 2013.

Card, D., R. Chetty, M. S. Feldstein, and E. Saez. 2010. Expanding access to administrative data for research in the United States. In *Ten Years and Beyond: Economists Answer NSF's Call for Long-Term Research Agendas*, C. L. Schultze and D. H. Newlon, eds. American Economic Association. Available at SSRN: http://ssrn.com/abstract=1888586 or http://dx.doi.org/10.2139/ssrn.1888586.

IDC. 2014, The Digital Universe of Opportunities: Rich Data and the Increasing Value of the Internet of Things. White Paper sponsored by EMC^2. http://www.emc.com/leadership/digital-universe/2014iview/index.htm?cmp=micro-big_data-general-emc&page=http%3A%2F%2Fwww.emc.com%2Fcampaign%2Fbigdata%2Findex.htm. Accessed: October 14, 2014.

Lee, C. A., ed. 2011. *I, Digital: Personal Collections in the Digital Era*. Chicago, IL: Society of American Archivists.

Lyman, P., and H. R. Varian. 2003. How Much Information? http://www2.sims.berkeley.edu/research/projects/how-much-info-2003/. Accessed June 16, 2013.

Manyika, J., M. Chui, B. Brown, J. Bughin, R. Dobbs, C. Roxburgh, and A. H. Byers. 2011. *Big Data: The Next Frontier for Innovation, Competition, and Productivity*. McKinsey Global Institute. http://www.mckinsey.com/insights/business_technology/big_data_the_next_frontier_for_innovation.

Office of Management and Budget. 2012. *Statistical Programs of the United States Government: Fiscal Year 2012*. http://www.whitehouse.gov/sites/default/files/omb/assets/information_and_regulatory_affairs/12statprog.pdf.

Spencer, N. 2012. How much data is created every minute? Blog post to *Visual News* posted on June 19. http://www.visualnews.com/2012/06/19/how-much-data-created-every-minute/?view=infographic. Accessed June 16, 2013.

Chapter 2

The Current State of Digital Curation

As the principles and practice of digital curation have developed, it has gained some recognition as a distinct field and garnered the attention of organizations dedicated to its improvement. Shared standards and norms for digital curation are appearing within many different disciplines and sectors and are filtering into other disciplines. Improved practices for ensuring the quality and durability of digital data are being established. The field of digital curation has many inducements to advance further, yet also faces some barriers. The benefits of doing digital curation are increasingly evident, but so are the actual and, often hidden, costs. In this chapter, benefits refer to measurable outcomes, such as the value of persistent access to high-quality and usable digital information products, as well as less tangible benefits, such as more complete and accurate data for decision making. Costs include the hardware, software, storage, and especially human labor of digital curation, as well as the potential costs to individuals, organizations, and society at large of failing to perform essential curatorial activities.

2.1 Evolution and Continuing Development of the Field of Digital Curation

Although processes, institutions, and skill sets for preserving and disseminating digital information have been known and in place in some disciplines for several decades, identifying that assemblage of practices is essential to fully establish the field of digital curation. Thus the field has grown from practices hardly recognized as curation per se—for example, researchers making note of metadata when collecting data—to international consortia engaged in defining shared norms and standards for digital curation.

The earliest organized efforts to curate machine-readable data in the United States began more than 50 years ago, when a few universities and government-supported research institutions established specialized repositories, called data archives or data libraries, to preserve and distribute numeric machine-readable data. The first social science data archive in the United States, the Roper Public Opinion Research Center, was established at Williams College in 1947. The first World Data Centers were created in 1957-1958 as an outgrowth of the International Geophysical Year. In 1962, the Inter-university Consortium for Political and Social Research (ICPSR) was founded at the University of Michigan.

During the latter part of the twentieth century, government agencies, national and state archives, and scientific data centers developed further capacity to manage, disseminate, and preserve machine-readable data. Academic libraries, corporate technical information centers, publishers, and others began to share responsibility for long-term curation of published literature in electronic journals and commercial databases. In the early 1990s, research collaboration among federal agencies such as the National Science Foundation (NSF), Defense Advanced Research Projects Agency, and National Aeronautics and Space Administration (NASA) and some public and private university libraries advanced information architecture, storage, and retrieval techniques. Libraries and cultural heritage institutions also launched large-scale efforts to digitize their collections and make them accessible to the public through the Internet (Waters and Garrett, 1996; National Research Council, 2000). In this period, much work in metadata, interoperation, shared resources, and discoverability was initiated, forming the foundation of digital library practices.

Organizations devoted explicitly to recognizing and developing the field of digital curation followed. For example, in 2002, the United Kingdom established the Digital Curation Centre (DCC).[1] In 2006, the *International Journal of Digital Curation*[2] was launched. In 2010, the National Digital Stewardship Alliance (NDSA) was established in the United States as a consortium of organizations committed to long-term preservation of digital information with particular emphasis on staffing needs and capacity building.[3] The United States also maintains a strong presence in international organizations with some complementary and growing interests in digital curation, such as the Research Data Alliance (RDA) and the Committee on Data for Science and Technology (CODATA).

The field of digital curation has established a good foundation. Many of the building blocks originated not only in digital repositories and libraries, scientific data centers, and topical archives, but from within the communities of researchers and producers of data themselves. In particular, the digital library community performed much of the foundational work in metadata, interoperation, shared resources, and discoverability. Development of the field of digital curation continues.

An early champion of the development of digital curation as a field was Jim Gray (although commentary on the topic had begun a number of years earlier, e.g., for the Human Genome Database (Fasman et al., 1997)). Gray articulated a "fourth paradigm" of scientific inquiry (Gray et al., 2002, 2005) to take its place beside the experimental, theoretical, and computational paradigms. This fourth paradigm, also known as "e-science," depends on data-driven methods—and thus on properly curated data. Particularly in "Online Scientific Data Curation, Publication, and Archiving," Gray and his collaborators (2002) working on the architecture for the Sloan Digital Sky Survey addressed digital curation as professional curators would, that is, as the critical work of annotation, preservation, and expert description of datasets. In doing so, Gray also affirmed that those performing curation would do well to learn, and not reinvent, the relevant concepts and techniques from librarians as they develop close collaboration with experts who have domain-specific expertise (Gray et al., 2002).

Today, the field of digital curation is challenged to keep pace with rapid change in all aspects of digital information. It must accommodate the immense increase in quantity of digital information, the many uses and reuses of that information, changing technology, and a continuum of people handling digital information in an array of organizational settings across all sectors. The continued advancement of digital curation involves not only meeting these challenges, but also maturing as a field. This currently includes the development and dissemination of norms and standards that enable sharing of digital information, best practices for improving and maintaining the quality of digital information, and skills and management techniques to protect and preserve digital information.

---

[1] See http://www.dcc.ac.uk.
[2] See http://www.ijdc.net.
[3] See http://www.digitalpreservation.gov/ndsa/.

2.2 Shared Norms and Standards

Any collection of digital information is apt to be utilized for a great variety of purposes by many different kinds of users in quite different settings. Any collection of digital information is also likely to be aggregated with other collections, and to be shared, accessed, analyzed, and stored using many different kinds of technology. All of this makes shared norms and standards in digital curation imperative. Such norms are emerging, though to different degrees in different contexts. This section reviews some recent efforts to establish norms and standards in digital curation, and then considers some of the current variation in adopting or establishing shared norms and standards for digital curation in government agencies, research communities, and business.

2.2.1 *Efforts to Establish Standards for Digital Curation*

The International Organization for Standardization (ISO), an independent nongovernmental membership organization that develops voluntary international standards, published the Open Archival Information System (OAIS) Reference Model (ISO 14721:2003) in 2003. The OAIS is a framework for the responsibilities, processes, and functions that any archiving organization needs in order to preserve information and provide long-term access. A set of auditing tools, the Trusted Repository Audit and Certification (TRAC), provides metrics for assessing conformance with the OAIS Reference Model. TRAC formed the basis for another ISO standard, the Trusted Digital Repository, which was adopted in 2012 (ISO 16363:2012). Although these standards do not prescribe a specific set of practices for digital curation, they do provide a common framework for identifying deficiencies in existing repositories and benchmarks against which to measure progress (e.g., Smith and Moore, 2007; Tarrant et al., 2009).

In addition to this international nongovernmental initiative, several national governments have either established some shared standards for digital curation or helped coordinate others' efforts to do so. In the United Kingdom, for example, the Digital Curation Centre (DCC) began in 2002 to provide guidance regarding standards and practices for digital curation and the skill sets and tools it requires. Over a decade later, through products such as the digital curation life-cycle model[4] and supporting documentation (frequently asked questions, checklists, etc.), the resource list for curation,[5] and the collection of training materials,[6] the DCC remains a valuable resource and exemplar for the emerging digital curation profession.

The United States, by contrast, lacks a single national center or association for digital curation. Many separate organizations are working to develop standards and define best practices for digital curation. Sometimes these organizations coordinate their efforts; too frequently they operate with little awareness of each other. Organizations such as the American Library Association's Association of College and Research Libraries' Digital Curation Interest Group, the Online Computer Library Center (now known as OCLC), the Coalition for Networked Information, the Library of Congress, the NDSA, and the Digital Library Federation, to name a few, raise awareness of digital curation, promote best practices, and provide professional development. Although some standards that support digital curation[7] have been adopted in the United States (e.g., ISO 19115-2 for geospatial metadata and Dublin Core for author, title, and

---

[4] See http://www.dcc.ac.uk/resources/curation-lifecycle-model.
[5] See http://www.dcc.ac.uk/resources.
[6] See http://www.dcc.ac.uk/training.
[7] See http://www.dcc.ac.uk/resources/metadata-standards/list.

subject tracking,[8]) many standards have not been implemented—even down to basic time encoding (ISO 8601). The specific reasons vary, but usually include absence of software implementations, inadequate documentation, or lack of an experienced community of peers from whom to seek guidance regarding the application of standards (Greenberg, 2009; Mize and Habermann, 2010).

2.2.2 Digital Curation Standards in Government Agencies

The degree to which government agencies adopt and follow standards for digital curation varies greatly. As noted in Chapter 1, the federal, state, and local governments create a massive flow of administrative documents as well as a large volume of statistical data for policy research. Government-sponsored scientific agencies also collect huge quantities of digital information. The curation of all this digital information is far from uniform. The extent to which scientific agencies have developed or adopted standards and policies for digital curation varies among agencies and between programs within a single agency.

Some of the best examples in the federal government include the three National Data Centers of the National Oceanic and Atmospheric Administration, the Earth Resources Observation and Science Center of the U.S. Geological Survey (USGS), and the National Center for Biotechnology Information of the National Institutes of Health (NIH). NASA established a federated set of data centers in the 1990s that collect, curate, and preserve data. Other agencies, such as the Department of Transportation, are now starting to develop repositories for their information assets.

Across government agencies that collect digital information and that fund scientific research, there has been an effort to draw attention to digital curation. Many have adopted policies that indicate a concern with the practice of proper curation of digital information. For instance, the NIH adopted a policy in 2003 that required all sponsored projects with more than $500,000 in direct costs to provide data management plans for sharing their data or to state why data sharing is not possible.[9] NSF similarly requires all grants submitted after January 18, 2011, to provide a data management plan. In February 2013, the White House Office of Science and Technology Policy (OSTP) issued a memo to heads of executive branch departments and agencies that asked them to submit plans intended to increase access to the results of federally funded scientific research, where *results* were defined to include peer-reviewed publications and data (Holdren, 2013). The federal government has been pressing hard for agencies to present properly curated . data for analysis with the creation of data.gov in 2009. Other high-level government reports (e.g., National Science Board, 2005; Interagency Working Group on Digital Data, 2009; Blue Ribbon Task Force on Sustainable Digital Preservation and Access, 2010) and publications by experts (e.g., Lord and Macdonald, 2003; Swan and Brown, 2008; Auckland, 2012; Lyon, 2012) have explored and illuminated digital curation and its many issues.

2.2.3 Digital Curation Standards in the Sciences

A number of changes are occurring in how researchers in the sciences aggregate, use, and share data. The gathering of original data has long been a hallmark of scientific research. The tradition of creating or collecting new data, analyzing the data, publishing the results, and often then abandoning the raw data is being displaced. The process of inquiry instead often begins with the selection of existing data followed by the application of new analytical

---

[8] See http://dublincore.org/.
[9] See http://grants1.nih.gov/grants/policy/data_sharing/data_sharing_guidance.htm.

techniques, such as data integration, data mining, text mining, modeling, simulation, and visualization, to those data. It concludes with new discoveries that are documented not only in publications, but also in derived or integrated data products, new analytical tools, and contributions of new data to reference datasets.

This new pattern of data usage has many implications (Parsons and Fox 2013). It puts a much greater imperative on the standardization of digital curation practices, so as to facilitate sharing. One of the best examples of this is the field of biodiversity informatics, a relatively new field that is devoted to the management and analysis of data concerning organisms, species, and biological communities. Its principal data types are:

- Natural history specimens in museums, herbaria, and other repositories;
- Observations of organisms in nature;
- Taxonomic names attached to these specimens and observations;
- Geographic distributions of species based on specimens and observations; and
- Character traits of organisms and species, including but not limited to morphology, behavior, and genetics.

As shown in Table 2-1, different data repositories relevant to the field of biodiversity informatics have utilized very different data standards.

Faced with this plethora of data standards, Biodiversity Informatics Standards (formerly, the Taxonomic Database Working Group, or TDWG[10]) undertook to develop standards to bring these databases into alignment. TDWG began operating in 1985 as an informal group of botanists interested in developing the ability to exchange data on taxonomic names and specimen data among botanical gardens and herbaria. The effort was formalized in 1994 as an activity of the International Union of Biological Sciences. Since that time, TDWG has collaborated with a variety of other professional groups to develop a range of ontologies and standards. Participants in these activities have come from taxonomy, evolutionary biology, computer science, philosophy, library information schools, and other fields. Together they have created a new community of practice devoted to data curation in biodiversity-related research. Perhaps the most important and far-reaching outcome of TDWG is the Darwin Core, a body of standards for data related to biological species, specimens, samples, observations, and events.[11]

---

[10] See http://www.tdwg.org/.

[11] A similar set of activities for genomic data began under the auspices of the Genomic Standards Consortium (GSC) in 2005. Representatives of TDWG and GSC are now engaged in an exploration of how the Darwin Core and genomic standards can be integrated. See http://www.gensc.org.

Table 2-1 Variety of Data Standards in Biodiversity Informatics

	Major Data Repositories	**Data Standards**
DNA and genes	International Nucleotide Sequence Database Collaboration[a] (GenBank, EMBL, DDBJ)	Gene Ontology[b] Genomic Standards Consortium[c]
Species	Global Biodiversity Information Facility[d]	Darwin Core Standards[e]
Populations	Global Population Dynamics Database[f]	Ecological Markup Language[g]
Ecological communities	DataONE[h]	Ecological Markup Language[i]
Coastal and marine environments	Digital Coast[j]	Coastal and Marine Ecological Classification Standard[k]
Habitats	European habitats[l]	EUNIS Habitat Classification[m]
Geospatial landscape data	U.S. National Geospatial Program[n]; European Environment Agency data/maps[o]	Open Geospatial Consortium[p]

[a] http://www.insdc.org/.
[b] http://www.geneontology.org/.
[c] http://gensc.org/gc_wiki/index.php/Main_Page.
[d] http://www.gbif.org.
[e] http://rs.tdwg.org/dwc/.
[f] http://www3.imperial.ac.uk/cpb/databases/gpdd.
[h] http://www.dataone.org/.
[i] http://knb.ecoinformatics.org/software/eml/.
[j] http://www.csc.noaa.gov/digitalcoast/data.
[k] http://www.csc.noaa.gov/digitalcoast/publications/cmecs.
[l] http://eunis.eea.europa.eu/related-reports.jsp.
[m] http://eunis.eea.europa.eu/habitats.jsp.
[n] http://www.usgs.gov/ngpo/.
[o] http://www.eea.europa.eu/data-and-maps.
[p] http://www.opengeospatial.org/standards/is.

Standardization of digital curation in biodiversity informatics has helped that field advance. The biodiversity network, VertNet,[12] is prospering. The *Biodiversity Data Journal* began accepting submissions of datasets in 2012. The Global Biodiversity Information Facility (GBIF)[13] is also well established and contributing to setting curatorial standards in the field of biodiversity informatics. GBIF is an intergovernmental organization created in 2001 after several years of planning conducted through the Organisation for Economic Co-operation and Development's Megascience Forum (now the Global Science Forum). GBIF provides a portal through which distributed databases of species and museum specimen information can be searched with a single query. National and regional node programs engage hundreds of data providers and assist them in complying with the standards for data transfer that are now established within the field of biodiversity informatics.

---

[12] See http://vertnet.org/.
[13] See http://www.gbif.org.

Although standards for sharing data and curating digital information are well developed in biodiversity informatics, they certainly are not unique to that field. Standards relating to the sharing and curating of data have also, for example, redefined the ecological sciences (Hunt et al., 2009), enabled major advances in astronomy (Goodman and Wong, 2009), and transformed research in the proteomics (see Box 2-1).

Box 2-1
Sharing Data in Proteomics

Over the last two decades, the biomedical sector has created several highly curated reference databases for genes, proteins, and other organic patterns. These are now considered an essential part of the life sciences research infrastructure. The Protein Data Bank (PDB) is one such resource. A PDB case study (Curry et al., 2010) observes that "Making available molecular representation, their 3-D coordinates and experimental data requires massive levels of curation to ensure that data inconsistencies are identified and corrected. A central data repository and sister sites accepts data in multiple formats, such as the legacy PDB format, the macromolecular Crystallographic Information File introduced in 1996, and the current Protein Data Bank Markup Language, valid since 2005" (pp. 42-43). The WorldWide Protein Data Bank uses a global hierarchical governance approach to data curation workflow. Its staff review and annotate each submitted entry prior to robotic curation checks for plausibility as part of the data deposition, processing, and distribution. "Distributing the curation workload across their sister sites helps to manage the activity" (Curry et al., 2010, p. 43).

Digital curation standards and practices among academic researchers are of course heterogeneous and vary greatly across fields. Fields such as genomics and proteomics have well-established standards for metadata and annotation (e.g., the Minimum Information About a Microarray Experiment and the Minimum Information About a Proteomics Experiment). Such fields also have mechanisms and incentives to reward researchers for producing, publishing, and sharing well-curated information, as well as specialists who support digital curation. Other fields, such as archaeology or psychology, are less advanced in building the standards, infrastructure, and knowledge bases necessary for digital curation in their domain. Even in fields where digital curation is less developed, expectations are changing about the availability and use of data. For example, more journals are requiring that data used as the basis of a published article be available for possible reanalysis by readers.

Another important shift in the sciences with implications for standards in digital curation is the increased relevance of access to digital information across disciplinary boundaries. Epidemiologists now track outbreaks of influenza using the very same data that marketers analyze to target consumers with advertising for flu remedies.[14] Literary scholars, computational linguists, and computer scientists all collaborate on crafting algorithms to mine vast bodies of digitized text, generating insights into the evolution of natural languages and the emergence of new concepts in popular culture. Some endeavors, such as the geospatial industry, arise exactly out of the collaboration across disciplines and sectors (see Box 2-2). This unprecedented sharing across not only areas of specialty, but across scholarly

---

[14] For a good example of the limitations of data reuse without a deep understanding of highly curated data, see Lazer et al. (2014).

> **Box 2-2**
> **The Geospatial Industry**
>
> According to Matteo Luccio in the Metacarta Blog, "The geospatial industry consists of individuals, private companies, nonprofit organizations, academic and research institutions, and government agencies that research, develop, manufacture, implement, and employ geospatial technology (also known as geomatics) and gather, store, integrate, manage, map, analyze, display, and distribute geographic information—i.e., information that is tied to a particular location on Earth"[15] (Luccio, 2008).
> The geospatial industry is using properly created digital information—from sources as dispersed as converted historic paper records and maps to remote sensing platforms registering the Earth's surface—to address questions as diverse as the introduction of invasive species, the spread of infectious diseases, and biological responses to climate change.

disciplines and even between those disciplines and businesses, governments, and nonprofit organizations also requires the articulation of and commitment to shared standards.

One of the finest illustrations of the datasets and analytical work that become possible when standards allow researchers to transcend the boundaries of disciplines, data platforms, formats, technologies, institutions, and even time is the International Comprehensive Ocean-Atmosphere Data Set (ICOADS). ICOADS is one of the most complete and heterogeneous collections of surface marine data in existence.[16] Begun in 1981, it has been built through a collaborative effort of U.S. and international scientific agencies. The immense curation effort involved in its construction included assembling 261 million individual marine reports from ships, buoys, and other platform types from 1662 to 2007, digitizing the earlier entries, standardizing and normalizing data, detecting and correcting errors, and performing quality control. The collection contains observations from many different observing systems that encompass the evolution of measurement technology over hundreds of years. ICOADS holds a wealth of data that can be used repeatedly for purposes and in ways simply unimaginable at the time of collection. Current uses include providing the base reference of ocean surface temperature, relative humidity, and sea level pressure for historical climate assessment and climate change modeling (Woodruff et. al., 2011). ICOADS has greatly leveraged both the short- and long-term value of the data to stakeholders across the research community and the public. It is also a testament to the maturation of the field of digital curation.

2.2.4 *Digital Curation Standards in the Business Sector*

In the business sector, as in the sciences, major changes in the use of digital information have led to greater appreciation and advancement of the field of digital curation. Increased reliance on and exploitation of digital information in businesses is a common and well-recognized phenomenon. Corporations utilize digital information to devise business strategies, inform decision making, manage inventories and production, and extend marketing. In the commercial sector, targeted online advertising and retail sales are driven by ever-richer data on consumer interests and behavior extracted from databases of "clicks" or selections. Entire new industries have also developed around producing and distributing

---

[15] See http://metacarta.wordpress.com.
[16] See http://icoads.noaa.gov/.

value-added information products. These include weather forecasts based on data from the National Weather Service[17] that are enhanced with special local or historical features; travel sites that aggregate air fares, hotel prices, local maps, and other destination information for travel planning and price comparison; or recommender systems that collect customer reviews for restaurants, stores, and services. Genealogical companies have acquired comprehensive census information on everyone in the United States born between 1790 and 1940 and everyone born in Britain between 1851 and 1911, have added billions of digitized records of births, deaths, and marriages, and have incorporated records on immigration and naturalization to create massive genealogical databases that subscribers can search.

In such a context, it has become understood that "digital curation is not just for curators," but rather, that it is a core business function. As in the sciences, clear norms and standards in digital curation can facilitate the work. As data become a commodity that is traded and sold, explicit standards and policies regarding security, access, privacy, intellectual property, and reuse of data have become essential elements of data stewardship, as evidenced in the July 2012 symposium organized by the study committee.[18] Professional standards for the design and operation of trusted repositories for digital information (ISO 14721:2003) require a governance structure, business and succession plans, designation of responsibility for data management and preservation, preservation policies, and mechanisms (Smith and Moore, 2007) to monitor and respond to changes in the technological, organizational, and financial environments (see the TRAC Audit and certification checklist[19]). Industry best practices also include the consideration of digital curation in acquisition plans for major systems or system upgrades and in decisions to contract with third-party service providers for storage and other services. Long-term data preservation methodologies are being integrated at executive levels across the entire enterprise (Rappa, 2012), and there has been growth in new positions such as chief data officer (Raskino, 2013).

Within the business sector, the motion picture industry may be exemplary for according explicit attention to digital curation—perhaps because its entire product line is now created, distributed, and stored in digital form. The Academy of Motion Picture Arts and Sciences' 2007 report, *The Digital Dilemma,* addressed archiving and accessing digital motion pictures from the perspective of the major motion picture studios. It highlighted two key findings:

- *Every enterprise has similar problems and issues with digital data preservation.*
- *No enterprise yet has a long-term strategy or solution that does not require significant and ongoing capital investment and operational expense.* (Academy of Motion Picture Arts and Sciences, 2007).

The report concluded that a cross-industry approach is required and that no single department or division could or should address digital curation challenges alone (see Box 2-3). Further, it suggested that collaboration must be fostered among stakeholders, especially to develop common strategies for digital preservation. The report also noted, "Archiving digital

---

[17] See http://www.weather.gov/.
[18] See http://sites.nationalacademies.org/PGA/brdi/PGA_070217.
[19] See http://www.crl.edu/content.asp?l1=13&l2=58&l3=162&l4=91.

Box 2-3
Seeking Standards for Digital Curation in the Motion Picture Industry

The Academy of Motion Picture Arts and Sciences' 2007 report, *The Digital Dilemma,* noted the need for collaboration in devising strategies and standards for digital curation. In particular: "The following questions raised by using digital technologies cannot be answered by any single department or division:
- *What is the value of the content?*
- *Who determines the value of the content?*
- *What content will be archived?*
- *Who determines what content will be archived?*
- *How will the content be archived?*
- *Who determines how the content will be archived?"* (Academy of Motion Picture Arts and Sciences, 2007).
---

data requires a more active management approach, and a more collaborative partnership among producers, archivists and users to exploit its full benefits."

While the initial report was from the perspective of the major motion picture studios, a subsequent report, *The Digital Dilemma 2* (Academy of Motion Picture Arts and Sciences, 2012), looked at all of the other producers of audiovisual content, such as nonprofits, independent filmmakers, documentarians who create far more content than the major studios (Maltz, 2012). Curation for audio and visual works in this category falls to university-based archives, independent nonprofit organizations, state archives, public libraries, museums, and independent private archives. These lack the funding, technical infrastructure, trained staff, and institutional support that the major studios have at their disposal for digital curation. Consequently, despite a desire for digitization, access, and preservation of material such as documentaries, oral histories, and interviews, only a small portion are being digitized and made accessible and an even smaller portion of born-digital recordings are captured, curated, and preserved. Nonetheless, the report noted an awareness of the significance of digital curation and of the importance of establishing policies and standards regarding digital curation, even among entities without the resources to implement them (Academy of Motion Picture Arts and Sciences, 2012).

2.3 Best Practices for Good Quality

As noted in Chapter 1, not all digital information is an asset. Enhancing information by cleaning data, detecting and correcting errors, and compiling comprehensive metadata can improve its quality and render it valuable. Good-quality digital information requires good-quality digital curation, and developing best practices to ensure good-quality data is another aspect of the maturation of the field of digital curation.

A recent NSF-funded workshop, "Curating for Quality: Ensuring Data Quality to Enable New Science" (Marchionini et al., 2012), yielded insights into the question of data quality and its relationship to digital curation. Quoting directly from the report's Executive Summary:

- *There are many perspectives on quality: quality assessment will depend on whether the agent making the assessment is a data curator, curation professional, or end user (including algorithms);*

- *Quality can be assessed based on technical, logical, semantic, or cultural criteria and issues; and*
- *Quality can be assessed at different granularities that include data item, data set, data collection, or disciplinary repository.*

The workshop also identified several key challenges (Marchionini et al., 2012):

- *Selection strategies—how to determine what is most valuable to preserve;*
- *How much and which context to include—how to ensure that data are interpretable and usable in the future, what metadata to include;*
- *Tools and techniques to support painless curation implementation and sharing, and technologies that apply across disciplines; and*
- *Cost and accountability models—how to balance selection, context decisions with cost constraints.*

While the NSF-funded workshop addressed data quality from the perspective of the sciences, other efforts have been made to consider data quality in ways appropriate to the business sector. Curry et al. (2010) identified the following aspects of data quality as most relevant for digital curation to address in the context of business use:

- Discoverability and Accessibility—Users can find and access data in a straightforward manner.
- Completeness—All of the requisite information is available.
- Interpretation—The meaning of the data is unambiguous.
- Accuracy—The data correctly represent the underlying parameters.
- Consistency—The data do not contradict themselves, and values are uniform.
- Provenance and Reputation—The data can be reliably tracked back to their original source, and the source of the data is legitimate (Buneman et al. 2001).
- Timeliness—The data are up to date for the task at hand.

2.4 Curating for Durability

Preserving digital information has become an increasing concern of digital curation as digital information is repeatedly reaccessed to address new questions and as the technology for handling digital information continues to change. The ephemeral nature of digital data poses significant challenges in all sectors. The scientific enterprise needs long-term storage and preservation of digital information, not only so research findings can be confirmed, but so that new methods can be used to analyze existing data, new hypotheses can be tested, and entirely unanticipated questions might be answered.

The Jet Propulsion Laboratory is cataloguing its primary collection of about 200,000 space mission tapes, wound on 12-inch reels and stored in airtight metal canisters. The tapes contain data relevant to long-term trends such as global climate change and tropical deforestation, but the tapes' greatest value, researchers say, may lie in the light they can shed on scientific questions that have not yet been posed. For example, NASA scientists ignored ozone data gathered on space flights in the 1970s because the readings were so low they thought they were erroneous. In the 1980s, after British scientists suggested that a dangerous thinning of the ozone layer was under way, NASA scientists were able to confirm the observation from the old data. Without tomorrow's context, we do not know what is valuable today. Preservation of data, well beyond the original context and purpose of its collection, is a paramount responsibility of curation (Duerr et al. 2004).

Even when that responsibility is embraced, changing technology raises many hurdles to fulfilling it. The cultural heritage sector faces the challenge of preserving its holdings quite starkly. The records and artifacts of cultural, social, and political heritage are increasingly captured in a wide range of media and formats that are not necessarily built with long-term preservation in mind, are subject to rapid deterioration, and depend on software and storage technologies that often become quickly obsolete (ACLS, 2006). This may be a familiar problem for earlier nondigital formats, such as audio recordings (see Box 2-4), but it is not exclusive to those formats.

Box 2-4
The Loss of Cultural Heritage Through Deterioration of Records and Technological Change

Sound recordings are a striking example of cultural heritage data at high risk of loss. These include music, oral histories, and radio broadcasts preserved in a wide variety of formats and media. According to a 2004 survey, an estimated 46 million sound recordings were held by public institutions (such as libraries and archives) in the United States, but the exact number—and more crucially, the condition of these recordings—was not known (Heritage Health Index, 2004). In fact, respondents to the survey noted that roughly 44 percent of those reported sound recordings were currently in "unknown" condition.

The fragility of these data was confirmed in 2010 in a study by the National Recording Preservation Board of the Library of Congress and the Council on Library Information and Resources (Bamberger and Brylawski, 2010). This study also concluded that many historical recordings already have been lost or cannot be accessed by the public. This lost data include many of the recordings of radio's first decade, from 1925 to 1935.

Other endangered records of cultural heritage include the radio programs of the Armed Forces Radio Service (AFRS), established by the U.S. government during World War II to provide radio programs to U.S. personnel around the world. Most AFRS programing was distributed regularly on transcription discs and included an edited version of commercial radio broadcasts as well as original programming by the AFRS. The Library of Congress holds more than 100,000 of these discs, making it the second largest collection of radio programs in its collection, and the single largest resource for the study of radio broadcast programming between 1942 and the early twenty-first century. Today, however, the AFRS no longer produces transcription discs and exclusively uses satellite technology to distribute its broadcasts, making it difficult for the Library of Congress—or any other institution—to collect and preserve this important cultural content (Bamberger and Brylawski, 2010).

Cultural records are increasingly created digitally (ACLS, 2006) and initially hosted or stored on e-mail servers, websites, blogs, and social media sites such as Flickr, YouTube, and Facebook. While the common assumption is that digital formats are less vulnerable to loss than older storage formats such as magnetic tapes, wax cylinders, or vinyl records, this is not necessarily the case. The risks to digital formats may be different, but are still substantial. Indeed, a 2010 study by the National Recording Preservation Board of the Library of Congress and the Council on Library Information and Resources (Bamberger and Brylawski, 2010) concluded that new digital audio recordings may actually be more fragile than those held in earlier formats and at risk of being lost even more quickly. This is because digital sound files can be easily corrupted, and widely used media, such as CD-ROM discs, may last only about 3 to 5 years before files start to degrade. To prevent the loss of cultural heritage and other cultural and research resources that result from changes in technology, digital curation attends to the persistence and durability of digital information, rather than the carriers or media on which they are stored. In the case of audio recordings, this involves audio experts actively monitoring, constantly maintaining, migrating, and frequently backing up the archives for which they are responsible.

2.5 Further Advancement in the Field of Digital Curation

The field of digital curation has evolved and matured. It now faces inducements to advance, yet also many impediments to further development. Some of these inducements and impediments are outlined here.

2.5.5 Inducements to the Advancement of Digital Curation

Among the surest inducements to improved digital curation are the scientific discoveries that have been made possible because of it, through the sophisticated use and reuse of properly curated digital information. Similarly, gains in the business sector that have been possible only through its myriad data-driven strategies and extensive digital assets provide a strong incentive to the further development of digital curation. A few other specific inducements to the advancement of digital curation as a field are reviewed here.

2.5.5.1 Inducement: Organizations

One impetus to the advancement of the field of digital curation is the existence of many organizations dedicated to the purpose of building the field, either generally or within a specific discipline. The DCC is an example of an organization whose efforts are to further the field of digital curation generally, while TDWG and GBIF, all discussed above, are facilitating the improvement of digital curation in their own disciplines. Numerous other organizations, universities, and data centers are developing guidance and training to help investigators manage their data effectively (e.g., DataONE, ICPSR,[20] and the Federation of Earth Science Information Partners[21]). The existence of such organizations does not guarantee advancement of the field, however. Both the patrons of these programs and the programs themselves could benefit from assessing their offerings and committing to more coordination to reduce overlap as well as gaps.

---

[20] See http://www.icpsr.umich.edu/icpsrweb/datamanagement/index.jsp.
[21] See http://wiki.esipfed.org/index.php/Data_Management_Workshop.

Nonetheless, these initiatives indicate a recognition of the value of advancing the field of digital curation and a willingness to devote organizational resources to that end.

2.5.5.2 Inducement: Government Requirements

Concern for digital curation is increasing across government agencies, resulting in requirements to perform digital curation as well as resources for the field. The National Archives and Records Administration (NARA), for example, is actively encouraging and assisting federal agencies to transition from paper to electronic records. As federal agencies receive approval to convert to digital records, the flow of those records considered important enough for temporary or permanent storage by NARA is increasing rapidly. As a result, NARA has been adding new job titles to deal with digital records: digital imaging specialist, dynamic media preservation specialist, and information technology specialist.[22]

As noted earlier, a number of government funding agencies, including NSF[23] and NIH,[24] have also instituted requirements for data management plans as a condition for applying for research support from the federal government. The Institute of Museum and Library Services (IMLS) is providing general terms and guidance for its grant applicants.[25] The purpose of the data management plans is to ensure that research results and data collected and produced with public funds are available to the public, to encourage individual investigators to assume responsibility for managing their information assets, to promote good data management practices, and to facilitate data sharing and reuse—all of which should help to advance the field of digital curation.

At the time of this report, the committee was unable to identify any published formal assessment of the NIH data management plan requirement on improvements in data management. In the case of NSF and IMLS, data management plan requirements were put into place only in the past 2 to 3 years, so it is premature to measure their impact on data management. Nevertheless, the NSF plan appears to have fostered awareness of the importance of digital curation of data, as evidenced by numerous reports and plans[26] and the emergence of tools to support the development of data management plans (e.g., DMPtool by the California Digital Library and DataONE).[27]

---

[22] See http://www.archives.gov/careers/jobs/positions.html (accessed 20 August, 2012).
[23] See http://www.nsf.gov/bfa/dias/policy/dmp.jsp.
[24] See http://grants.nih.gov/grants/policy/data_sharing/.
[25] See http://www.imls.gov/applicants/projects_that_develop_digital_content.aspx.
[26] See http://www.arl.org/focus-areas/e-research/data-access-management-and-sharing/nsf-data-sharing-policy#.VRvPTxaNw5h.
[27] See https://dmp.cdlib.org/.

2.5.5.3 Inducement: Protection of Assets

While the proper curation of digital information assets is of unquestioned value for the business sector, this extends beyond the digital information assets that businesses themselves control. At the BRDI Symposium on Digital Curation in the Era of Big Data: Career Opportunities and Educational Requirements, Steve Miller of IBM noted that, with the rise of social media, the general public is becoming the curator of many companies' reputational assets, affecting their ongoing reputation and performance. As a result, companies find themselves struggling with curation issues for information that they do not own, but that is about them (e.g., Wikipedia,[28] Yelp[29]) and may be factually incorrect or even fabricated (Miller, 2012).

2.5.5.4 Inducement: Professional Recognition

Explicit recognition of the work done by trained digital curators may also be an impetus to development of the field. That recognition comes in many forms. Citation practices are beginning to credit curators. Citation recommendations for the Federation of Earth Science Information Partners, for example, help "data stewards define and maintain precise, persistent citations for data they manage and provide fair credit for data creators or authors, data stewards, and other critical people in the data production and curation process."[30] Formal job titles for digital curators, whether in the research, government, or private sectors, are also bringing recognition to the field.

2.5.5.5 Inducement: Openness and Transparency

Another impetus to the continued development of digital curation as a field is a public concern for consistency and transparency in organizations of all types. In part, this concern is reflected in governments' commitment to enable broad access to digital information to a wide range of researchers and to allow the private sector and individual citizens to benefit from research findings and data funded through tax revenues. This is viewed as an essential means to stimulate further scientific discovery and maximize return on publicly funded investments in research. For example, the Obama administration issued its Open Government Directive in December 2009, indicating that providing and maintaining data from all U.S. federal agencies in digital and accessible form is a central priority (Orszag, 2009). This commitment was underscored in February 2013 when the White House Office of Science and Technology Policy issued the aforementioned policy directive on public access to research information and data (Holdren, 2013). That directive instructed U.S. federal agencies to create plans for maximizing the accessibility of digital data and results from federally funded research, noting that increased accessibility facilitates productive reuse of data. The directive states that "policies that mobilize data for re-use through long-term preservation and broader public access" significantly enhance the impact of the federal investment in research. Greater access or reuse of these resources also "accelerates scientific breakthroughs and innovation, promotes entrepreneurship, and enhances economic growth and job creation." The directive covers data and information resulting from research in all disciplines. Federal agencies that generate or fund a significant amount of cultural heritage data—including the Smithsonian Institution, the National Endowment for the

---

[28] See http://www.wikipedia.org.
[29] See http://www.yelp.com.
[30] See http://wiki.esipfed.org/index.php/Interagency_Data_Stewardship/Citations/provider_guidelines.

Humanities, and the Institute of Museum and Library Services—are included in its scope.

Other research funders also now seek to ensure openness and transparency through digital curation policies that make the outputs of funded research, including data, readily accessible. Such policies are proliferating nationally and internationally, driven by the perceived potential that greater, long-term accessibility of data can increase the return on the initial investment in research (Houghton, 2011; Zuniga and Wunsch-Vincent, 2012, Beagrie and Houghton, 2014). The Organisation for Economic Co-operation and Development (OECD) Principles and Guidelines illustrate this approach, noting: "Sharing and open access to publicly funded research data not only helps maximize the research potential of new digital technologies and networks, but provides greater returns from the public investment in research" (OECD, 2007). In recent years, research sponsors have recognized that communication of results is an essential, inextricable part of the research process and have explicitly earmarked funds to cover some of the costs of dissemination (e.g., the Wellcome Trust Open Access Policy[31] was released in 2006 and strengthened in 2012 [Wellcome Trust, 2012]).

The value of consistent, transparent, and open practices of digital curation is not limited to the return on investments in research. Such practices may also assuage public concerns about privacy, confidentiality, and security. Several high-profile events, such as exposure of the National Security Agency's PRISM program that captures metadata on private e-mail and telephone messages, security and data breaches at large commercial firms such as Target, and the electronic recordkeeping challenges associated with the Affordable Care Act, have all exacerbated this concern. Transparent practices and shared standards in the field of digital curation could help allay public concern.

2.6 Impediments to the Advancement of the Field of Digital Curation

Although the gains that will result from the continued development of the field of digital curation seem sufficient to compel that trajectory, nonetheless there are barriers to continued advancement. Some of the policies and practices discussed above regarding adherence to standards and sharing of digital information may not be universally welcomed. Researchers may be reluctant to share data for fear of being scooped or be unwilling to invest time and effort to curate their own data when the reward system recognizes publication of results more than publication or deposit of data (Hedstrom and Niu, 2008; Koch, 2009). Many disciplines have not developed a culture of data sharing (Borgman, 2010; Tenopir et al., 2011; Thessen and Patterson, 2011), and the forms of data that researchers are most willing to release are not necessarily fit for new applications (Cragin et al., 2010). In the private sector, data are often tightly held because they are a resource that provides a competitive advantage (Maltz, 2012; Miller, 2012). While public concerns might lead to more open data and transparency, misuse of digital information and real or perceived invasions of privacy could also result in a backlash against the use of digital information. Statutory requirements and social norms for confidentiality and privacy could thwart the prospect of using many types of personally identifiable information for research (Dwork, 2007). Best practices and effective techniques for balancing openness and privacy protection might enhance the value of digital curation as a field.

Beyond these potential barriers briefly sketched here, one other merits further discussion. That is the lack of financial resources. A lack of funding for digital curation will threaten not

---

[31] Position Statement in Support of Open and Unrestricted Access to Published Research, http://www.wellcome.ac.uk/About-us/Policy/Spotlight-issues/Open-access/Policy/index.htm.

only the digital information that curatorial practices are meant to protect and preserve, but will also threaten the development of the field, for example, through an insufficiency of training programs to prepare a properly skilled workforce. Concerns over available resources—particularly monetary constraints—are fairly commonplace (Tenopir et al., 2011). Systematic investments in digital curation, including educating and training a workforce capable of tackling the complex challenges of the field, have been proposed to accelerate the transition to data-intensive science (Hey et al., 2009) and to a data-driven economy, although the nature of the latter is still poorly understood or measured (Mandel, 2012).

An example of acutely insufficient financial resources for digital curation is the National Biological Information Infrastructure (NBII; USGS, 2011). The NBII was begun in 1994 to establish a single, online source of biological resource data and information from vetted sources. In 2001, the NBII established regional and thematic nodes on the web, providing access to the information resources most important to their individual geographic or scientific niches. The NBII provided visible benefits to the biological resource community, enabling data owners to maintain critical assets that might not otherwise be made broadly available and providing a single, web-based source of data from numerous organizations to facilitate search. It also provided users with direct access to data resources deeply embedded in structured databases relevant to biology that might not be accessible via a nonspecialized search engine (USGS, 2011).

Despite its demonstrated value to the research community, the NBII (like many federally funded projects) suffered a series of deep budget cuts. Funding for NBII fell from a high of $7 million in fiscal year (FY) 2010 to $3.8 million in FY 2011, and to $0 in FY 2012, resulting in its mandated termination. As a result, the main NBII website (originally at http://wdc.nbii.gov/ma), along with all of its associated nodes, was taken offline in January 2012. The public face of these data was effectively lost. Behind the scenes, though, the USGS and the Socio-Economic Data Application Center[32] staff at Columbia University have been working with partners to identify ways—to the extent possible—to try to fill the gap left by the loss of the NBII program and to further curate the data from the decades-long, multimillion-dollar investment by taxpayers, scientists, and the federal government for widespread use.

2.7 Measuring the Benefits of Digital Curation

The value of active management and enhancement of digital information assets for current and future use may be evident in the abstract, but further efforts to measure the benefits and costs of digital curation are also part of assessing the current state of the field. This section considers aspects of measuring the benefits of digital curation. The following section will address the measuring of costs.

Benefits from digital curation accrue in a variety of ways: through efficiency gains, reductions in operating costs, opportunities to create value by doing things in new ways, and opportunities to market new products or offer new services. Precisely how much value curation adds to digital information and for whom are impossible to measure in the absence of an explicit market for digital products with differential pricing for curated and uncurated data. A further impediment to measuring the benefit of digital curation activities is that these activities do not map neatly to specific job titles or occupations with known ranges for prevailing salaries or wages. Moreover, the workforce demands for people with digital curation knowledge and skills

---

[32] See http://sedac.ciesin.columbia.edu/.

and the nature of the tasks they will be expected to perform will be contingent on the level of investments that organizations make in automating digital curation processes. The benefits of digital curation thus include many unknowns that defy estimation. Nonetheless, a framework may be constructed, defining different dimensions along which benefits of digital curation can be identified, examined, and eventually measured.

The work of Beagrie et al. (2010) provides one such framework. They propose a high-level taxonomy consisting of three dimensions (Outcome, Time Frame, and Beneficiary) along which the benefits of digital curation may be placed (see Figure 2-1). The outcomes of digital curation fall along a spectrum from direct through indirect. The time frame in which benefits are realized could range from near term to long term. The beneficiaries of digital curation could be internal to the curating entity or external, such as to a funding sponsor, another organization entirely, or some combination of them. Different metrics might be developed for measuring the types of outcome (direct and indirect), the time frame in which the benefit is realized, and the type of beneficiary (stakeholders internal to or affiliated with the organization undertaking the curation activity, and stakeholders external to or not affiliated with the organization undertaking the data curation activity). This is not an exhaustive list of dimensions; other dimensions could be added. The framework is a mnemonic to help ensure that no benefits are overlooked when assessing digital curation.

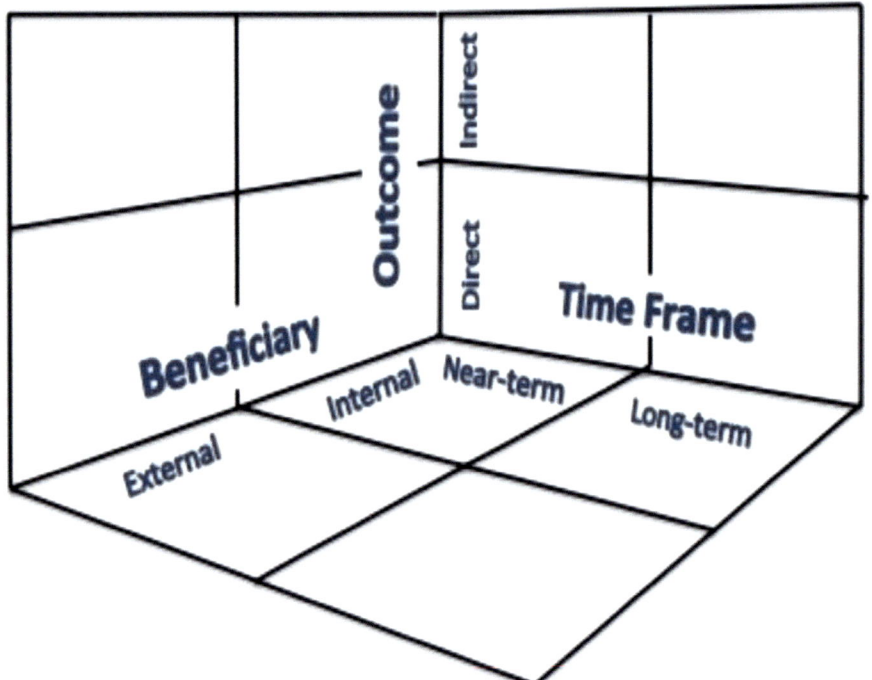

Figure 2-1 Framework for measuring the benefits of digital curation. Recast of the Beagrie et al. (2010) taxonomy into three dimensions (rather than sides on a triangle). In the process of assessing a digital curation project, benefits would be identified and located in each cell.

Fry et al. (2008) also address how to measure the benefits of curating and sharing research data. In their view, cost savings are the most direct and directly measurable benefit of digital curation. These savings can accrue to the depositor on the input side or the user on the output side. Fry et al. (2008) also emphasize that cost savings can accrue to multiple participants in the research process. Research funders and research institutions may realize savings through increased efficiencies in collecting data, greater reuse of existing data, and elimination of redundant effort. Researchers themselves may benefit by their directing effort away from collecting, calibrating, and cleaning new data toward analysis of large pools of available digital resources and the development of new analytical techniques. If high-quality data were available for reuse, the burden placed on less obvious participants in the research process, such as the large numbers of human subjects required for experiments and the large number of respondents needed to produce valid samples for surveys, might be reduced. Authors, reviewers, editors, and publishers of research papers may also benefit from the ready availability of high-quality data to review and substantiate research findings.

Fry et al. (2008) identify other significant benefits in addition to direct cost savings that can accrue from digital curation. These include:

- Increased collaboration and cost sharing;
- Greater use of data in teaching and research training, especially in graduate and postgraduate education projects;
- New opportunities and uses for data, including data mining;
- Creation of a more complete and transparent record of research;
- Improved research evaluation and the direction of research; and
- Creation of new areas of research, new industries, or new support services.

Other benefits might be added to this list, including improved interdisciplinary research and greater opportunities for innovation as data sources not previously available are leveraged. Fry and colleagues have also begun to create a methodology for quantifying the financial value of the benefits of digital curation. That work is in its preliminary stages.

2.8 Measuring the Costs of Digital Curation

Measuring the costs of digital curation involves many variables and some assumptions, but efforts are being made to devise metrics. Guidelines are available to identify cost factors that should be included in any cost-benefit analysis. The guidelines provided by Beagrie et al. (2008) in the Keeping Research Data Safe (KRDS) project permit the identification of a large percentage of the digital curation costs of a repository in a research environment. The KRDS model also provides a suite of useful tools, including a detailed description of cost variables and units, and an activity model—known as KRDS2—for identifying research data activities outside the repository, with cost implications for digital curation.

The KRDS model addresses costs in both the prearchive and archive phases of digital curation. The prearchive phase includes activities related to creating research data. In this phase, implications for repository costs are considered and strategies for data collection and creation are designed and implemented. The archive phase consists of the acquisition or disposal, ingestion, storage, and management of data, as well as access and user support. Administrative overhead, support services, common services, and estate costs must be included for both phases. The KRDS2 activity model provides a life-cycle costing method that identifies measurable component activities, variable cost drivers, and resources (e.g., staff time and equipment). It also provides a useful mechanism for understanding the costs of activities (Beagrie et al., 2010).

Using case studies from a variety of disciplines, Beagrie and colleagues (2010) observed the following cost trends associated with long-term curation:

- Acquisition and ingestion activities consistently cost the most.
- The costs of archival storage and preservation activities are consistently a small proportion of the overall costs and are significantly lower than the costs of acquisition, ingestion, or accession activities in all case studies.
- Potential cost-efficiencies can come from development of tools to support automation of ingestion and accession activities for curation and preservation.
- The costs of long-term data curation and preservation are dominated by fixed costs that vary little with the size of the collections.
- Staff members are the major cost component overall, and there is a minimum base level of staff coverage, skills, and equipment required for any service.
- Fixed-cost activities can reduce the per-unit cost of long-term preservation by leveraging economies of scale, using multi-institutional collaboration, and outsourcing as appropriate.

They further noted a trend of relatively high preservation costs in the early years of establishing a particular data collection. These are likely to be reduced substantially over time for longer-lived data collections (Beagrie et al., 2010).

For addressing the concerns of this report, that is, estimating demand for a digital curation workforce and providing sufficient and appropriate training for that workforce, it is highly significant that the KRDS2 model identifies staff costs as the major cost component overall. As the cost of digital storage decreases over time, activities that require human expertise and manual intervention will constitute an even greater portion of curation costs. It is also important to recognize that costs associated with staffing—including investment in the development of the right people, with the right expertise, at the right time—are interdependent and shifting.

Further, a number of strategies exist to reduce these staffing costs, including outsourcing, crowdsourcing, and the economies of scale achieved through collaboration across a community of researchers or among many research institutions. A key conclusion in the Beagrie et al. (2010) cost model is that leveraging such strategies could reduce the per-unit costs of digital curation. Other models helpful for analyzing the costs of digital curation include the Total Cost of Preservation model developed by the University of California Curation Center and the California Digital Library, as well as the Cost Model for Digital Preservation, the LIFE (Lifecycle Information for E-Literature) model, and the DataSpace model. These models focus primarily on the preservation aspect of digital curation. Recent work by Beagrie and Houghton (2014) integrates elements from a number of available cost models and demonstrates the value of synthesizing a range of quantitative approaches.

Two limitations of the existing cost models are worth noting. First, because the models estimate costs from the perspective of the repository, costs borne by the creators of digital information are not included in the model. When the creators of digital information have the foresight, resources, and ability to conform to digital curation standards and best practices, the costs to the repository for acquisition, ingestion, and quality control can be reduced. The extent of digital curation activity that occurs in this phase is easily underestimated. This is particularly so when the producers of digital information are performing the curation activities, such as adding metadata, keeping track of versions of records, conducting quality assurance, and annotating entries. A strong dependency exists between the extent, quality, and timing of digital curation activities conducted by producers and the eventual costs of curation to a repository. If producers fail to provide contextual information about the content, structure, format, quality, and other aspects of their data, curators and other repository staff must spend time and effort to recover or add that contextual information (Niu and Hedstrom, 2007).

Second, although prearchive curation activities are highly desirable from the perspective of the repository because they accelerate the movement of digital information into preservation infrastructure and can lower costs to a repository, the benefits of prearchive curation activities to data producers are widely variable and not well articulated. Compelling examples of coordinated curation across the entire information life cycle that mutually benefit the producers, stewards, and consumers of digital information could serve as powerful models for the types of coordinated efforts envisioned in this report.

While proper digital curation imposes costs, inadequate digital curation can lead to many risks and even greater costs. Organizations of all types incur costs when they are unable to deliver the right information, in the right form, to the right players, at the right time. For some organizations, investments in digital curation are a form of risk mitigation. Commercial enterprises that collect revenue directly from sale or distribution of data and information products or services need effective digital curation for quality assurance and financial viability. Digital curation can help organizations that depend on access to accurate, timely, and authoritative digital information for business intelligence and decision making to reduce costly mistakes from the strategic level to operations.

The oil spill at British Petroleum's (BP's) *Deepwater Horizon* oil platform in the Gulf of Mexico in 2010 provides an example of the risks of inadequate digital curation. Real-time data about the causes of the spill, its extent and impact, and the efficacy of efforts to cap the oil well were critical to addressing the situation and organizing a response that would limit damage to marine life, water quality, shoreline and beaches, and recreational activities in the Gulf of Mexico. BP, numerous federal agencies (the U.S. Coast Guard, NOAA, the Department of Energy, the USGS, and others), as well as state and local agencies provided data on the potential environmental and economic consequences of the spill. Disputes over the accuracy of the data provided by BP surfaced quickly and persisted throughout the containment and cleanup efforts. This not only hampered efforts to limit the disaster itself, but also resulted in damage to BP's reputation and the assessment of fines and penalties on the company. Subsequent multibillion-dollar liability lawsuits against BP included one brought by the U.S. Department of Justice on behalf of multiple U.S. agencies that was settled in early 2012, with BP agreeing to fines of $4.5 billion (Krauss and Schwartz, 2012).

Inadequate investment in or performance of digital curation can also be costly in public policy. Insufficient and inefficient management of digital assets can become an issue in policy debates. Accurate and accessible information is critical to defining policy options, evaluating the potential effects of different options, and developing methods for implementing directives and statutes that have far-reaching immediate and long-term consequences for the public.

The importance of high-quality information for policy making was recognized during the health care reform process in the United States. The Agency for Healthcare Research and Quality (AHRQ) worked to ensure that the policy makers considering sweeping changes to U.S. health care statutes were presented with accurate, up-to-date, and easily accessible data. Such data came from myriad sources and covered all aspects of the health care system, including insurance coverage, care delivery, and outcomes. To meet these extremely broad needs, AHRQ convened a planning group to assess its data provision capabilities and to develop a strategy to optimize the availability of information and data for enactment and implementation of health care reform (AHRQ, 2009). The planning group recognized the need for a long-term effort to collect and analyze new data. It decided, however, that it was more efficient in the short term to focus on making current data more available, linking existing data resources, and in some cases identifying strategies to increase the timeliness of a subset of high-priority data. The planning group emphasized the immediate need for AHRQ to develop a "stand-ready" capacity to provide data that would optimize the effectiveness of the policy-making process, while also noting the ongoing need for data to track the impact of existing or new policies. This range of strategies was meant to mitigate the risks to health policy reform incurred as a result of poorly curated data.

Daunting as it is to estimate the actual costs of digital curation, estimating the costs of not doing digital curation is even harder. And costly as digital curation may be, the cost of failing to undertake these activities may be even greater. If irreplaceable digital information assets are lost, destroyed, or become inaccessible or uninterpretable through inadequate or improper curation practices, how can that loss be quantified? Unfortunately, such losses abound. In the cultural heritage sector, this includes many historic audio recordings and radio broadcasts, as discussed in Section 2.4. In the sciences, the wealth of information contained in the NBII, also discussed above, is at grave risk of being lost. Had not the Magellan Tapes (see Section 2.4) been properly curated, questions regarding ozone depletion could not have been fully addressed, even though such questions had not even been posed when the tapes were made. This would have been a costly loss. The potential for compounded costs due to the failure to conduct proper curation, and the difficulty of estimating those costs, are perhaps best illustrated by ICOADS, discussed in Section 2.2.3. Who, in 1662, conducting a cost-benefit analysis of the value of maintaining and storing ships' logs, could have foreseen their use, once digitized, by researchers in the twenty-first century?

2.9   Conclusions and Recommendations

Conclusion 2.1: Demands for readily accessible, accurate, useful, and usable digital information from researchers, information-intensive industries, and consumers have exposed limitations, vulnerabilities, and missed opportunities for science, business, and government, as a result of the immaturity and ad hoc nature of digital curation. There is also a push for greater openness and transparency across many sectors of society. Taken together, these factors are creating an urgent need for policies, services, technologies, and expertise in digital curation. Although the benefits of digital curation are poorly understood and not well articulated, significant opportunities exist to embed digital curation deeply into an organization's practices to reduce costs and increase benefits.

Conclusion 2.2: There are many inducements that could drive advances in the field of digital curation:

- Organizations that can serve as leaders, models, and sources of good curation practices;
- Government requirements for managing, sharing, and archiving information in digital form;
- Protection of digital assets to build trust and satisfy consumers and to maintain competitiveness in business and scientific research;

- Rewards and professional recognition for the value that curation adds to digital information; and
- Pressure from consumers, citizens, and society at large for accountability and transparency in business and government.

Conclusion 2.3: There are also barriers to developing the capacity for comprehensive, affordable, and effective digital curation. Some impediments, such as attitudes about sharing data and concerns over privacy, competitive advantage, security, and misuse of digital information, are difficult to delineate or measure. Insufficient financial resources for digital curation are a commonplace concern.

Conclusion 2.4: Cost models and studies of digital curation costs consistently identify human resources as the most costly component of digital curation. Current cost models are likely to underestimate the costs of curation tasks performed by the creators and producers of digital information, because no techniques have been developed to segregate or measure curation costs prior to accessioning into a repository. There is a pressing need to identify, segregate, and measure the costs of curation tasks that are embedded in scientific research and common business processes.

Conclusion 2.5: Although standards and good practices for digital curation are emerging, there is great variability in the extent to which standards and effective practices are being adopted within scientific disciplines, commercial enterprises, and government agencies. The absence of coordination across different sectors of the economy and different organizations has led to limited adoption of consistent standards for digital curation and resulted in the fragmented dissemination of good practices.

Recommendation 2.1: Organizations across multiple sectors of the economy should create inducements for and lower barriers to digital curation. The Office of Science and Technology Policy should lead policy development and prioritize strategic resource investments for digital curation. Leaders in information-intensive industries should advocate for the benefits of digital curation for product innovation, competitiveness, reputation management, and consumer satisfaction. Leaders of scientific organizations and professional societies should promote mechanisms for recognition and rewards for scientific and professional contributions to digital curation.

Recommendation 2.2: Research communities, government agencies, commercial firms, and educational institutions should work together to accelerate the development and adoption of digital curation standards and good practices. This includes (1) the development and promotion of standards for meaningful exchange of digital information across disciplinary and organizational boundaries and (2) interoperability between systems used to collect, accumulate, and analyze digital information and the repositories, data centers, cloud services, and other providers with long-term stewardship and dissemination responsibilities.

Recommendation 2.3: Researchers in economics, business analysis, process design, workflow, and curation should collaborate to identify, estimate or measure, and predict costs associated with digital curation. The National Science Foundation, the Institute of Museum and Library Services, relevant foundations, and industry groups should solicit proposals for and fund such research.

Recommendation 2.4: Scientific and professional organizations, advocacy groups, and private-sector entities should articulate, explain, and measure the benefits derived from digital curation, including "after-market" benefits, risk mitigation, and opportunities for private-sector investment, innovation, and development of curation technologies and services. The benefits should include outcomes that generate measurable value, as well as less tangible benefits such as the accessibility of digital information over time for scientific research, organizational learning, long-term trend analysis, policy impact analysis, and even personal entertainment. Such research is necessary for the development and testing of sophisticated cost-benefit (or cost-value) models and metrics that encompass the full range of digital curation activities in many types of organizations.

2.10 References

Academy of Motion Picture Arts and Sciences, Science and Technology Council. 2007. *The Digital Dilemma: Strategic Issues in Archiving and Accessing Digital Motion Picture Materials.* http://www.oscars.org/science-technology/council/projects/digitaldilemma/.

Academy of Motion Picture Arts and Sciences, Science and Technology Council. 2012. *The Digital Dilemma 2: Perspectives from Independent Filmmakers, Documentarians and Nonprofit Audiovisual Archives.* http://www.oscars.org/science-technology/sci-tech-projects/digital-dilemma-2.

ACLS (American Council of Learned Societies). 2006. Our Cultural Commonwealth—Report of the ACLS Commission on Cyberinfrastructure for Humanities and Social Sciences. http://www.acls.org/cyberinfrastructure/ourculturalcommonwealth.pdf. Accessed September 8, 2013.

AHRQ (Agency for Healthcare Research and Quality). 2009. Filling the Information Needs for Healthcare Reform. Expert Meeting Summary and Identification of Next Steps. March 19. http://www.ahrq.gov/research/data/hinfosum.html. Accessed September 26, 2013.

Auckland, M. 2012. *Re-skilling for Research: An Investigation into the Role and Skills of Subject and Liaison Librarians Required to Effectively Support the Evolving Information Needs of Researchers.* Research Libraries UK. http://www.rluk.ac.uk/files/RLUK%20Re-skilling.pdf.

Bamberger, R,, and S, Brylawski. 2010. *The State of Recorded Sound Preservation in the United States: A National Legacy at Risk in the Digital Age.* Washington, DC: Council on Library and Information Resources and Library of Congress. http://www.clir.org/pubs/reports/pub148/reports/pub148/pub148.pdf.

Beagrie, N., and J. W. Houghton. 2014. The Value and Impact of Data Sharing and Curation: A Synthesis of Three Recent Studies of UK Research Data Centres. Joint Information Systems Committee Report. . http://repository.jisc.ac.uk/5568/1/iDF308_-_Digital_Infrastructure_Directions_Report%2C_Jan14_v1-04.pdf.

Beagrie, N., J. Chruszcz, and B. Lavoie. 2008. *Keeping Research Data Safe: A Cost Model and Guidance for UK Universities.* Higher Education Funding Council for England. http://www.webarchive.org.uk/wayback/archive/20140615221657/http://www.jisc.ac.uk/media/documents/publications/keepingresearchdatasafe0408.pdf.

Beagrie, N., B. F. Lavoie, and M. Woollard. 2010. *Keeping Research Data Safe2*. Higher Education Funding Council for England. http://www.jisc.ac.uk/publications/reports/2010/keepingresearchdatasafe2.aspx#downloads.

Blue Ribbon Task Force on Sustainable Digital Preservation and Access. 2010. *Sustainable Economics for a Digital Planet: Ensuring Long-Term Access to Digital Information.* http://brtf.sdsc.edu/.

Buneman, P., S. Khanna, and W.-C. Tan. 2001. Why and where: A characterization of data provenance. Pp. 316-330 in *Database Theory—ICDT 2001*, J. Van den Bussche and V. Vianu, eds. Berlin, Heidelberg: Springer, doi:10.1007/3-540-44503-X_20.

Borgman, C.L. 2010. Research data: Who will share what, with whom, when, and why? Retrieved from: http://works.bepress.com/borgman/238 and http://www.nlc.gov.cn/yjfw/zm/index_en.html.

Cragin, M. H., C. L. Palmer, J. R. Carlson, and M. Witt. 2010. Data sharing, small science and institutional repositories. *Philosophical Transactions of the Royal Society A: Mathematical, Physical and Engineering Sciences* 368(1926):4023-4038.

Curry, E., A. Freitas, and S. O'Riain. 2010. The role of community-driven data curation for enterprises. Pp. 25-47 in *Linking Enterprise Data*, D. Wood, ed.. Springer.

Duerr, R., M. A. Parsons, R. Weaver, and J. Beitler. 2004. The International Polar Year: Making data available for the long-term. In *Proceedings of the Fall American Geophysical Union Conference*. San Francisco, CA, December. ftp://sidads.colorado.edu/pub/ppp/conf_ppp/Duerr/The_International_Polar_Year:_Making_Data_and_Information_Available_for_the_Long_Term.ppt. Accessed February 2, 2013.

Dwork, C. 2007. Ask a better question, get a better answer: A new approach to private data analysis. Pp. 18-27 in *Proceedings of 2007 International Conference on Database Theory,* . T. Schwentick and D. Suciu, eds. Berlin, Heidelberg: Springer-Verlag. http://research.microsoft.com/pubs/64345/icdt.pdf.

Fasman, K. H., S. I. Letovsky, P. Li, R. W. Cottingham, and D. T. Kingsbury. 1997. The GDB™ Human Genome Database Anno 1997. *Nucleic Acids Research* 25(1):72-80.

Fry, J., S. Lockyer, C. Oppenheim, J. Houghton, and B. Rasmussen. 2008. *Identifying Benefits Arising from the Curation and Open Sharing of Research Data Produced by UK Higher Education and Research Institutes*. Leicestershire, UK: Loughborough University. https://dspace.lboro.ac.uk/2134/4600.

Goodman, A. A., and C. G. Wong. 2009. Bringing the night sky closer: Discoveries in the data deluge. Pp. 39-44 in *The Fourth Paradigm: Data Intensive Scientific Discovery*, T. Hey, S. Tansley, and K. Tolle, eds. Redmond, WA: Microsoft External Research.

Gray, J., A. S. Szalay, A. R. Thakar, and C. Stoughton. 2002. Online scientific data curation, publication, and archiving. Pp. 103-107 in *SPIE Astronomical Telescopes and Instrumentation*. International Society for Optics and Photonics.

Gray, J., D. T. Liu, M. Nieto-Santisteban, A. Szalay, D. J. DeWitt, and G. Heber. 2005. Scientific data management in the coming decade. *ACM SIGMOD Record* 34(4):34-41.

Greenberg, J. 2009. Theoretical considerations of lifecycle modeling: An analysis of the Dryad Repository demonstrating automatic metadata propagation, inheritance, and value system adoption. *Cataloging & Classification Quarterly* 47(3-4):380-402.

Greenberg, J. 2012. Alfred P. Sloan Foundation perspective. Presented to the Symposium on Digital Curation in the Era of Big Data: Career Opportunities and Educational Requirements, Board on Research Data and Information, National Research Council, Washington, DC, July 19.

Hedstrom, M., and J. Niu. 2008. Incentives to create "archive-ready" data: Implications for archives and records management. In *Proceedings of the American Archivist (SAA) Research Forum*, San Francisco, CA, August 26-30.

Heritage Health Index. 2004. *A Public Trust at Risk: The Heritage Health Index Report on the State of Americas Collections*. Washington, DC: Heritage Preservation, Inc. http://www.imls.gov/assets/1/assetmanager/hhifull.pdf.

Hey, T., S. Tansley, and K. Tolle, eds. 2009. *The Fourth Paradigm: Data Intensive Scientific Discovery*. Redmond, WA: Microsoft Research.

Holdren, J. P. 2013. Increasing Access to the Results of Federally Funded Scientific Research. Memorandum for the Heads of Executive Departments and Agencies. Office of Science and Technology Policy. http://www.whitehouse.gov/sites/default/files/microsites/ostp/ostp_public_access_memo_2013.pdf.

Houghton, J. W. 2011. *Costs and Benefits of Data Provision*. Report to the Australian National Data Service. http://ands.org.au/resource/houghton-cost-benefit-study.pdf. Accessed September 8, 2013.

Houghton, J. W., B. Rasmussen, and P. J. Sheehan. 2010. *Economic and Social Returns on Investment in Open Archiving Publicly Funded Research Outputs*. Report to SPARC by Victoria University's Centre for Strategic Economic Studies. http://www.cfses.com/FRPAA/. Accessed September 8, 2013.

Hunt, J. R., D. D. Baldocchi, and C. van Ingen. 2009. Redefining ecological science using data. Pp 21-26 in *The Fourth Paradigm: Data Intensive Scientific Discovery*, T. Hey, S. Tansley, and K. Tolle, eds. Redmond, WA: Microsoft Research.

Interagency Working Group on Digital Data. 2009. *Harnessing the Power of Digital Data for Science and Society*. http://www.nitrd.gov/About/Harnessing_Power_Web.pdf.

Koch, S. 2009. Personal open science challenges. Steve Kock Science blog. http://stevekochscience.blogspot.com/2009/02/personal-open-science-challenges.html.

Krauss, C., and J. Schwartz. 2012. BP will plead guilty and pay over $4 billion. *The New York Times*, November 15. http://www.nytimes.com/2012/11/16/business/global/16iht-bp16.html?pagewanted=all.

Lazer, D., R. Kennedy, G. King, and A. Vespignani. 2014. The parable of Google flu: Traps in big data analysis. *Science* 343(6176):1203-1205. http://j.mp/1ii4ETo.

Lord, P., and A. Macdonald. 2003. *e-Science Curation Report: Data Curation for e-Science in the UK: An Audit to Establish Requirements for Future Curation and Provision*. Digital Archiving Consultancy Limited. http://www.jisc.ac.uk/uploaded_documents/e-ScienceReportFinal.pdf.

Luccio, M. 2008. What is the geospatial industry? *MetaCarta Blog*, August 7. http://metacarta.wordpress.com/2008/08/07/what-is-the-geospatial-industry/. Accessed July 8, 2013.

Lyon, L. 2007. *Dealing with Data: Roles, Rights, Responsibilities and Relationships*. Consultancy Report. UKOLN, University of Bath. http://www.webarchive.org.uk/wayback/archive/20140615031414/http://www.jisc.ac.uk/media/documents/programmes/digitalrepositories/dealing_with_data_report-final.pdf.

Lyon, L. 2012. The informatics transform: Re-engineering libraries for the data decade. *International Journal of Digital Curation* 7(1):126-138. http://ijdc.net/index.php/ijdc/article/view/210/279.

Maltz, A. 2012. Entertainment industry perspective. Presented to the Symposium on Digital Curation in the Era of Big Data: Career Opportunities and Educational Requirements, Board on Research Data and Information, National Research Council, Washington, DC, July 19.

Mandel, M. 2012. Beyond Goods and Services: The (Unmeasured) Rise of the Data-Driven Economy. Policy Memo from the Progressive Policy Institute. http://www.progressivepolicy.org/wp-content/uploads/2012/10/10.2012-Mandel_Beyond-Goods-and-Services_The-Unmeasured-Rise-of-the-Data-Driven-Economy.pdf. Accessed April 19, 2013.

Marchionini, G., C. A. Lee, H. Bowden, and M. Lesk. 2012. *Curating for Quality: Ensuring Data Quality to Enable New Science.* Arlington, VA: National Science Foundation. http://openscholar.mit.edu/sites/default/files/dept/files/altman2012-mitigating_threats_to_data_quality_throughout_the_curation_lifecycle.pdf. Accessed April 16, 2013.

Miller, S. 2012. The future of work. Presented to the Symposium on Digital Curation in the Era of Big Data: Career Opportunities and Educational Requirements, Board on Research Data and Information, National Research Council, Washington, DC, July 19.

Mize, J., and R. T. Habermann. 2010. Automating metadata for dynamic datasets. In *OCEANS 2010*, September 20-23. Institute of Electrical and Electronic Engineers. . http://ieeexplore.ieee.org/stamp/stamp.jsp?tp=&arnumber=5663837&isnumber=5663781.

National Research Council. 2000. *LC21: A Digital Strategy for the Library of Congress.* Washington, DC: The National Academies Press. http://www.nap.edu/catalog.php?record_id=9940.

National Science Board. 2005. *Long-Lived Digital Data Collections: Enabling Research and Education in the 21st Century.* Washington, DC: National Science Foundation. http://www.nsf.gov/pubs/2005/nsb0540/.

Niu, J., and M. Hedstrom. 2007. Streamlining the "producer/archive" interface: Mechanisms to reduce delays in ingest and release of social science data. Presented at DigCCurr2007 Conference, Chapel Hill, NC, April 18-20.

OECD (Organisation for Economic Co-operation and Development). 2007. OECD Principles and Guidelines for Access to Research Data from Public Funding. http://www.oecd.org/science/scienceandtechnologypolicy/38500813.pdf.

Orszag, P. R. 2009. Open Government Directive. Memorandum for the Heads of Executive Departments and Agencies. Office of Management and Budget, Executive Office of the President. December 8. http://www.whitehouse.gov/sites/default/files/omb/assets/memoranda_2010/m-10-06.pdf.

Parsons, M. A., and P. A. Fox. 2013. Is data publication the right metaphor? *Data Science Journal* 12:WDS32-WDS46. http://dx.doi.org/10.2481/dsj.WDS-042.

Smith, M., and R. W. Moore. 2007. Digital archive policies and trusted digital repositories.*International Journal of Digital Curation* 2(1):92-101.

Swan, A., and S. Brown. 2008. *Skills, Role & Career Structure of Data Scientists and Curators: An Assessment of Current Practice and Future Needs.* Bristol, UK: JISC. http://www.jisc.ac.uk/publications/reports/2008/dataskillscareersfinalreport.aspx.

Tarrant, D., B. O'Steen, T. Brody, S. Hitchcock, N. Jefferies, and L. Carr. 2009. Using OAI-ORE to transform digital repositories into interoperable storage and services applications. *Code{4}lib Journal* (6). http://journal.code4lib.org/articles/1062.

Tenopir, C., S. Allard, K. Douglass, A. U. Aydinoglu, L. Wu, E. Read, M. Manoff, and M. Frame. 2011. Data sharing by scientists: Practices and perceptions. *PLoS ONE* 6(6): e21101. http://www.plosone.org/article/info:doi/10.1371/journal.pone.0021101.

Thessen, A E., and D J. Patterson. 2011. Data issues in the life sciences. *ZooKeys* 150:15.

USGS (U.S. Geological Survey). 2011. NBII to be taken offline permanently in January. *Access* 14(3). http://www.usgs.gov/core_science_systems/Access/p1111-1.html.

Waters, D., and J. Garrett. 1996. *Preserving Digital Information. Report of the Task Force on Archiving of Digital Information.* Washington, DC: Commission on Preservation and

Access. http://www.oclc.org/content/dam/research/activities/digpresstudy/final-report.pdf. Accessed March 1, 2013.

Wellcome Trust. 2012. Wellcome Trust strengthens its open access policy. Press Release, June 28. . http://www.wellcome.ac.uk/News/Media-office/Press-releases/2012/WTVM055745.htm.

Woodruff, S. D., Worley, S. J., Lubker, S. J., Ji, Z., Eric Freeman, J., Berry, D. I., and Wilkinson, C. 2011. ICOADS Release 2.5: extensions and enhancements to the surface marine meteorological archive. *International Journal of Climatology*, *31*(7), 951-967.

Zuniga, P., and S. Wunsch-Vincent. 2012. Harnessing the benefits of publicly funded research. *WIPO Magazine* June. Accessed September 8, 2013. http://www.wipo.int/wipo_magazine/en/2012/03/article_0008.html.

Chapter 3

Current and Future Demand for a Digital Curation Workforce

Digital curation is in demand and will be in demand across many sectors, from scientific research to business, health care, and cultural expression. Current demand is difficult to ascertain precisely, because of the dispersed nature of digital curation activities undertaken by many different actors in a broad range of organizational settings and institutional contexts. A lack of basic occupational data covering the field of digital curation also obstructs estimates. Nonetheless, estimates can be made of demand for both the current and future workforce in digital curation. This chapter explores trends in current and future demand and then considers some factors that may affect future demand, particularly automation.

3.1 Difficulties in Estimating Current Demand

Tracking current demand for the workforce in digital curation is difficult for three reasons. One is that the job is dispersed. As discussed in Chapter 1, digital curation is best conceived as a continuum, a set activities that may be accomplished by one or many people, whose professional identification and training may relate primarily to curation — or hardly at all. It may occur in a dedicated data repository, or in very different research, business, or cultural settings. How is a job to be counted, when it exists in so many permutations?

A second factor confounding estimates of the workforce in digital curation is that this is a fairly new field. The job title "digital curator" is only just emerging. Most federal agencies, for example, continue to describe their jobs dealing with the retention, organization, and dissemination of federal records as records managers, a job title that reflects the designation from an earlier time when records were kept as paper documents. Records manager remains an important federal government job title, even as the duties have expanded to deal with digital records. Identifying which jobs, though traditionally identified perhaps as records manager, data analyst, or librarian, have now evolved into de facto digital curator is a challenge.

The third obstacle to estimating current demand for the workforce in digital curation is, ironically, the lack of data. The primary source of statistics on employment is the federal government, principally the Bureau of Labor Statistics (BLS). At this time, the BLS does not track digital curation as a separate occupation.

All three of these factors impede the estimation of current demand for the workforce in digital curation. All three need to be addressed. Regarding the first: conceiving of digital curation as a continuum of activities is an accurate way to capture how the field is actually practiced, but it may be taken to an extreme. When digital curation is defined as *the active management and enhancement of the utility of digital information assets for current and future use*—then almost all activities associated with information can be said to be digital curation. A more reasonable account of the continuum of activities might place curation-related jobs in order, from those in which digital curation is the sole activity of a job's incumbent, to those in which digital curation occurs from time-to-time in a job that is embedded in some other domain. Job data may be sifted with this understanding of the continuum in mind.

The second impediment to estimating the demand for digital curators, that is, the lack of a job with that title, is beginning to recede. For example, a recent opening for the job of data curator and analyst identified by Miller (2012) included the following duties:

- Develop and maintain tools/codes for day-to-day data extraction, curation, and management;
- Extract and provide clean data;
- Measure and track quality improvements in data;
- Increase awareness of value of the data quality;
- Work closely with IT groups and statistical teams; and
- Create and present summary statistics and reports

This job title is clearly for a digital curation occupation as the committee has defined it. Although the title of digital curator has not existed in the past, it is beginning to emerge. This will facilitate the tracking of demand.

The third obstacle to estimating demand—the lack of government statistics—merits further attention. The chief source of data on jobs in the United States is the BLS. Unfortunately, none of the jobs in the BLS Standard Occupational Classification (SOC) is titled "Digital Curation." This may change as the SOC is revised.

To count the number of workers in an occupation, the BLS uses as its employment measure the number of jobs in a particular occupation, not the number of workers, and defines job growth as the sum of the new jobs created and the number of new hires of replacement workers needed due to turnover. The BLS gathers employment and wage data through its Occupational Employment Statistics (OES), a semiannual survey of approximately 200,000 firms from the 50 states and 4 territories. The BLS sample is selected from about 7 million firms. The self-employed are not included in the survey.

Under the OES program, new occupations are added as they emerge. The expansion of occupations that are listed in the SOC system and that are surveyed and analyzed reflects major changes in the U.S. labor force. The SOC was last updated in 2010, when 24 new detailed occupations and codes were added. Four of these were in computers and mathematical occupations, consisting of (1) information security analysts, (2) web developers, (3) computer network architects, and (4) computer network support specialists (BLS, 2012a).

The next revision of the SOC will be completed in 2018. The revision is likely to add further computer and mathematical occupations, but it is not clear at this point that digital curation and related occupations will be included. The process for developing a new SOC includes a solicitation for public recommendations of new SOC codes. A *Federal Register* (BLS, 2014) notice soliciting recommendations for new occupations was issued on May 22, 2014. The public was encouraged to respond to the notice and to make recommendations and provide justifications for new occupations that should be added to the SOC. Digital curation responsibilities currently incorporated into other occupations could be separated in 2018 or in a later revision. Thereafter, revisions are planned to be made every 10 years (BLS 2012a).

3.2 Estimating Current Demand: Job Openings

While estimating current demand for a digital curation workforce is difficult, it is not impossible. Knowledge of how digital curation is conducted and how occupations are labeled can help in interpreting the job data that are available. Private-sector sources can supplement incomplete or inadequate government data. One very useful source of information is data on job openings.

The data used here come from job postings over the past 7 years, using historical data from the private company Indeed.com, a recruitment advertising network that aggregates job listings worldwide and across the private and public sectors. Indeed.com gathers listings of job openings from thousands of sites, including job boards, individual company career pages, newspaper classified sections, business and professional associations, and blogs. The company's listings cover the private sector extensively, and also gather openings from federal, state, and local governments. Some details of the methods used by Indeed.com are available on the company's website.[1]

Table 3-1 compares a number of traditional computer and mathematical occupations to job titles related to digital curation. As expected, some traditional computer and mathematical occupations have many more job openings than those in most job titles related to digital curation. Yet, despite their small absolute size, most job openings related to digital curation have been growing much more rapidly than the openings for computer and mathematical occupations.

All of the computer and mathematical occupations in Table 3-1 had more openings in 2012 than they did in 2005. The growth rate of these occupations was substantial, but none of the occupations grew by more than 120 percent during the 7-year period. Several of these individual occupations have a substantial share of the job openings compared to all occupations, with two occupations having more than 1 percent of economy-wide job openings—software developers (1.4 percent) and computer systems analysts (1.1 percent).

---

[1] http://www.indeed.com/.

Table 3-1 Job Trends from Indeed.com. Job Openings in Computer and Mathematical Occupations and in Occupations Related to Digital Curation, 2005 to 2012.

Occupation	Percent of Openings 2005	Percent of Openings 2012
Computer and mathematical occupations		
Software developers	0.8	1.4
computer systems analysts	0.8	1.1
computer support specialists	0.1	0.3
network administrators	0.1	0.2
computer programmers	0.12	0.15
computer research scientists	0.05	0.08
data base administrators	0.01	0.03
Occupations Related to Digital Curation		
Enterprise architects	0.3	0.8
data governance	0.1	0.4
enterprise governance architects	0.01	0.08
data stewards	0.01	0.05
information curators	0.001	0.01
Data Curators	0.0	0.002
Librarians	0.06	0.06
Archivists	0.03	0.06

NOTE: Percentages are rounded.
SOURCE: Data from www.indeed.com/jobtrends. Accessed August 17, 2012.

By contrast, job openings related to digital curation have experienced much larger increases in percentage terms, with all but the librarian occupation at least doubling over the past 7 years. The librarian occupation experienced no growth, although tasks requiring digital curation expertise may nonetheless have increased within that occupation. Despite the rapid growth in job openings related to digital curation, most of these occupations have had few openings recently. Enterprise architect (0.8 percent) and data governance (0.4 percent) occupations, however, have had more job openings than four of the computer and mathematical occupations listed.

Data regarding job openings listed by Indeed.com also furnish the information presented in Figure 3-1. This figure illustrates the trend of an increase in job postings that seek digital curators from 2005 to 2012.

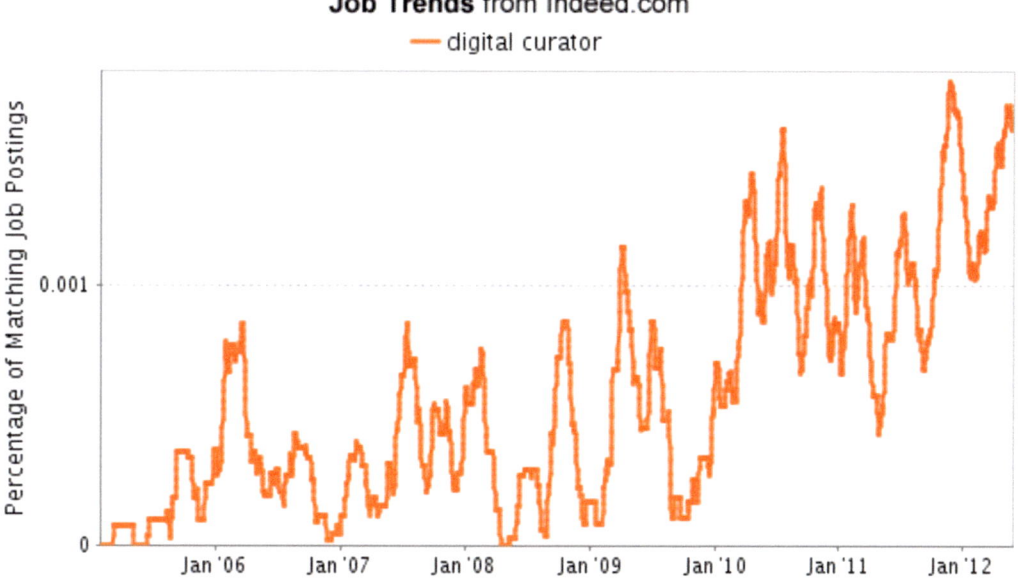

Figure 3-1 Job openings containing a "digital curator" job title, 2005 to 2012. Percentage of job openings found by Indeed.com that contain the term "digital curator."
SOURCE: Indeed.com (2012).

Figure 3-2 Trends in posting for positions that used the term "information steward" in the job description.
SOURCE: Data extracted from Indeed.com (2012).

Similarly, Figure 3.2 uses data from Indeed.com to illustrate the increase in postings for positions that used the term "information steward" in the job description. The colored arrow represents the upward trend, determined by the low point, the high point, and an estimate of the mean.

Figure 3-3 summarizes these parameters for a collection of 30 terms related to digital curation in one logarithmic chart. The arrow colors relate to the trend—green indicates a substantial upward trend, yellow a flat or more gradual change, and red indicates a significant downward trend.

One can observe from this figure that application areas such as social media, data analytics, and electronic medical records are experiencing substantial rates of growth, as are some of the core digital curation areas such as "data governance," "data steward," "digital repository," and "digital preservation." It is also worth noting that of these 30 terms, a third of them began to appear in position descriptions only after 2006.

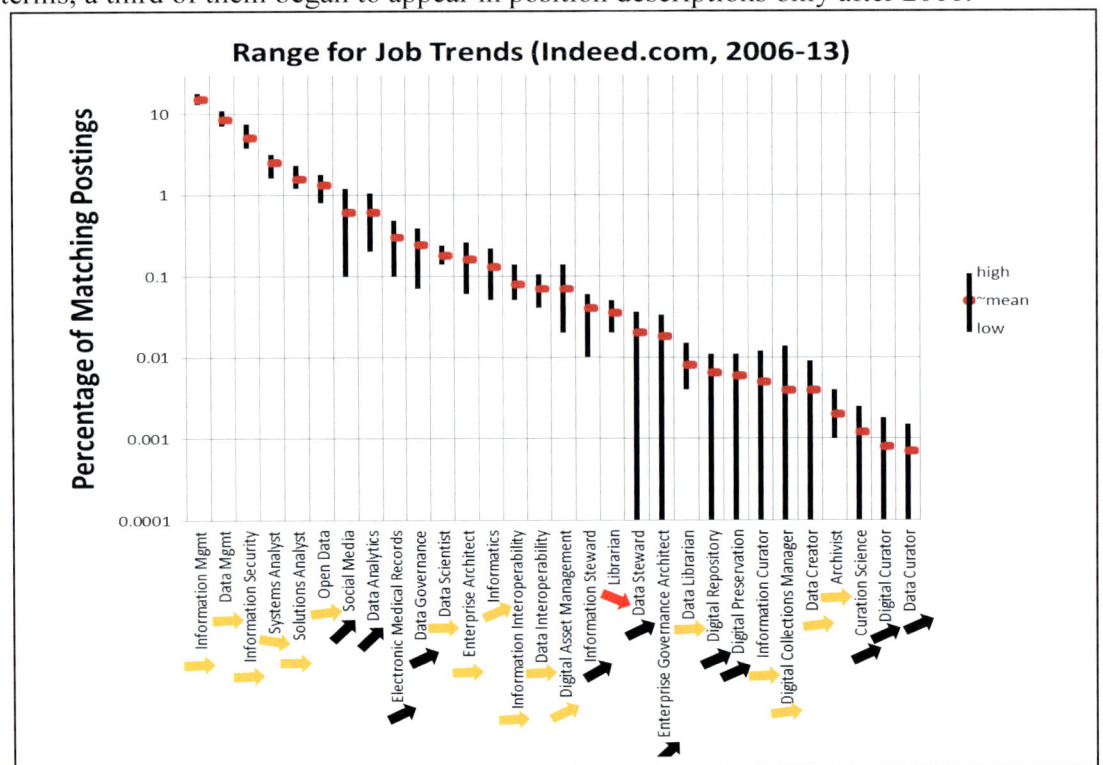

Figure 3-3 Percentage range and trend of jobs that contain identified terms. Green indicates a substantial upward trend, yellow a flat or more gradual change, and red a significant downward trend.
SOURCE: Indeed.com (2012).

A new study by the Conference Board identifies librarians, archivists, and curators as occupations that are likely to experience labor shortages in the next 10 years (Levanon et al., 2014). That study used government data and estimates of retirements by baby boomers, projected demand for employment, current unemployment and labor participation rates, percentage of positions in an occupation filled by immigrants, and projected productivity increases in 266 industries and 464 occupations in the U.S. The report found that most occupations with a high risk of labor shortages are projected to either experience high employment growth or have few new entrants to replace those who leave, but not both. For librarians, archivists, and curators the report projects an average employment growth of 8 percent over the decade, but an inadequate supply of new entrants to replace those who retire or leave the occupation (Levanon et al., 2014, p. 21). The study does not explicitly mention the implications of digital curation for the employment outlook for librarians, archivists, and curators, but it is worth noting that labor markets for occupations with knowledge and skills that are essential for digital curation, such as computer-related occupations, also are projected to be tight.

3.3   Estimating Current Demand: Placements

Other sources of information provide data helpful for estimating current demand for the workforce in digital curation. Students are now graduating from library and information science (LIS) schools, information schools (iSchools), and some management information systems programs in business schools with credentials in information management that fit the committee's definition of digital curation. The ease with which these recent graduates are placed in jobs and the initial trajectories of their careers provide indirect indicators of current market demand. The trends revealed by these indirect indicators are consistent with the trends presented above using data on job openings.

Interpreting these indirect data requires attention to the current state of professional training. Because digital curation is an emerging field, relevant professional training may not be reflected in the formal titles of final degrees. Rather, it may be found in the various programs, subspecialties, and certificates associated with professional degrees in library science, information science, and even business.

Despite the lack of public access to comprehensive survey data, some trends in the experience of LIS graduates can be discerned.[2] A 2011 survey of recent graduates of LIS programs in the United States, with respondents from 35 percent of alumni ($N = 2,162$) from 41 programs, support a possible trend in digital curation positions: "LIS programs and graduates alike reported an increase in the number of emerging job titles, including ... digital curator." Graduates also reported new responsibilities in digitization, data collection, and analytics (Maatta, 2012).

A survey of 2010 graduates in LIS by *Library Journal* showed that 78 percent were employed (Maatta, 2011). The titles for the digital curation–related positions accepted by graduates included:

- Data management specialist,
- Data management librarian,
- Data curator,
- Data librarian,
- Data services specialist,
- Data research scientist,
- Data curation specialist,
- *E*-science librarian,

---

[2] The Association for Library and Information Science Education (ALISE) compiles information on LIS graduates and their placement, but these data are available only to ALISE member organizations.

- *E*-research librarian,
- Science data librarian, and
- Research data librarian.

However, of the 1,346 respondents to this survey, only 80 (about 6 percent) listed positions explicitly related to digital curation (automation systems, database management, electronic or digital services, information technology), though many other positions may involve digital curation activities.

Information on the career directions of recent graduates from Syracuse University's Certificate of Advanced Study (CAS) in Data Science is also available. Launched in 2008, the CAS is a 15-credit graduate certificate that may be pursued either as a specialization within a master's degree or as a stand-alone program. The following information (Elizabeth Liddy, 2012) although numbers are small, illustrates the range of career paths that are becoming open to graduates with credentials in digital curation. Within 3 months of graduating, six of the eight graduates had accepted employment or entered graduate study as:

- A science research support librarian at a university-based research center;
- A digital curation librarian at a large public university;
- A metadata librarian at a small private college;
- A data coordinator at a private consulting and design company;
- A data, network, and translational research librarian at a university-based health sciences center; and
- A medical school student at a private university.

The University of Illinois at Urbana-Champaign also began a specialization in data curation within its library and information science master's degree program in 2007, with 63 graduates completing the specialization as of December 2012. Placement data for 83 percent of these graduates also confirms that curation professionals are filling positions in many types of organizations in a range of roles. Just under half have taken positions in academic settings in libraries at colleges and university across the country. The next largest group has been placed in the corporate sector, followed by a segment working in museums and other cultural heritage institutes, national data centers, and digital humanities and scientific research institutes. The more than 50 position titles are suggestive of how curation jobs are being formalized, ranging from data curator, data management consultant, research data librarian, and digital preservation librarian to data analyst, digital asset manager, and information architect (Palmer et al., 2014). A few graduates from Indiana University's School of Informatics have been placed in explicitly data-oriented positions with titles such as data analytics consultant and database and systems manager, with a range of other types of positions, many of which are designated software and systems developers and designers (Fox, 2012).

Data on placement from students who have completed certificate programs and professional LIS degrees is incomplete. Even less is known about the impacts of continuing education programs, short courses, and workshops on participants' careers. Further, digital curation education and training programs are so new that while some data on initial job placements for graduates are available, data on long-term career trajectories are lacking. More information on placements will contribute to estimates of demand for the workforce in digital curation.

3.4 Estimating Future Demand: Government Statistics

The effort to estimate future demand for the workforce in digital curation is again stymied by the lack of complete government statistics. However, some insight can be gained by studying the BLS data that are available. BLS makes occupational projections for over 700 job categories and 300 industries every 2 years, for the upcoming 10 years. As discussed in Section 3.1, none of the jobs in the BLS SOC is titled "Digital Curation." As a result, future demand for the digital curation workforce can only be approximated by examining employment projections for computer occupations and a few other occupations that are strongly related to digital curation.

BLS provides occupational outlook information to the general public and to analysts; the latest *Occupational Outlook Handbook* was published in March 2012, based on the period from 2010 to 2020. Data used for the occupational projections are gathered for 22 major occupational groups and 749 occupations based on the SOC. Digital curation is likely to be concentrated in four major occupational groups: (1) Business and Financial Operations (Matrix Code 13-0000); (2) Computer and Mathematical (15-0000); (3) Life, Physical, and Social Sciences (19-0000); and (4) Education, Training, and Library (25-0000).

These four major occupational groups are expected to experience above-average growth (see Table 3-2). The fastest growth in employment (22 percent) and the highest median wage ($73,720) is projected for Computer and Mathematical occupations, but Education, Training, and Library occupations (10.60 million) will employ more workers, as will Business and Financial Operations (7.96 million) (Lockard and Wolf, 2012). These data do not directly indicate a growth in digital curation jobs, but do suggest that growth will occur.

Table 3-2 Employment and Wages of Four Major Occupational Groups and All Occupational Groups, 2010 and Projected 2020

Occupation	Employment 2010 (millions)	Employment 2020 (millions)	Projected change (%)	Median Annual Wage ($, May 2010)
Business & financial operations	6.79	7.96	17.3	60,670
Computer & mathematical	3.54	4.32	22.0	73,720
Life, physical, and social science	1.23	1.42	15.5	58,530
Education, training, & library	9.19	10.60	15.3	45,690
Total, all occupations	143.07	163.54	14.3	33,840

SOURCE: Data from Lockard and Wolf (2012).

Table 3-3 Employment Data by Occupation, 2010, and Projections for 2010-2020

Occupation	Jobs 2010 (thousands)	Employment Change 2010-2020 (%)	Employment Change 2010-2020 (thousands)	Median Annual Wage 2011 ($)
Computer and mathematical occupations				
Computer support specialists	607.1	18.1	110.0	47,660
Computer systems analysts	544.4	22.1	120.4	78,770
Software developers, applications	520.8	27.6	143.8	89,280
Software developers, systems software	392.3	32.4	127.2	96,600
Computer programmers	363.1	12.0	43.7	72,630
Network and computer systems administrators	347.2	27.8	96.6	70,970
Computer and information systems managers	307.9	18.1	55.8	118,010
Information security specialists, web developers, and computer network architects	302.3	21.7	65.7	77,970
Database administrators	110.8	30.6	33.9	75,190

CURRENT AND FUTURE DEMAND FOR A DIGITAL CURATION WORKFORCE

Computer and information research scientists	28.2	18.7	5.4	101,080
Digital curation-related occupations				
Librarians	156.1	6.9	10.8	55,300
Archivists	6.1	11.7	0.7	46,750

SOURCES: Data from BLS (2012b); Csorny (2012).

Note that each of these four major occupational groups is growing more rapidly than the aggregate of all occupations. They also have higher median wages than the aggregate of all occupations.

Table 3.3 presents BLS projections for the various categories of computer and mathematical occupations, many of which may incorporate digital curation functions, as well as for librarians and archivists.

Many of the categories of computer and mathematical occupations employ large numbers of people, and many are expected to grow by greater than the 14.3 percent national average of all occupations between 2010 and 2020. All of these occupations have median wages of greater than the U.S. median wage of $33,840, and all but one have wages at least twice the median wage. The occupations of librarian and archivist, both related to digital curation, have a smaller number of jobs. Their projected rates of growth between 2010 and 2020 are less than average, while their median annual wage is greater than average but less than twice the median wages for all occupations.

Estimating demand for the workforce in digital curation is difficult, whether for the current period or in the future. Given the nature of digital curation activities—occurring along a continuum, and conducted by many different kinds of workers holding many different job titles and undertaking those activities either as primary professional focus or as a very limited subset of their jobs—the difficulty is not surprising. The lack of direct statistics is also a challenge. Nonetheless, estimates may be made, and they reveal a clear increase in the demand for a skilled workforce in digital curation.

3.5 Automation and Future Demand

The scale of future demand for a workforce in digital curation will be affected by many factors. Some of these—the overall health and growth of the economy, governments' increased requirements for digital curation, the continued transfer of historic records to digitized formats—need no further reflection here. One other factor invites more consideration. That is the impact that automation of curatorial tasks may have on the demand for human capital in digital curation.

Automation is already integral to digital curation, in all disciplines, domains, and sectors. Most digital curation involves some combination of human decisions and manual effort as well as automated tools and processes. How much of curation is automated and how that automation is integrated with manual tasks along the whole continuum of the workflows of curation varies a great deal. Variability in the automation of digital curation tasks derives from many sources, including the size and resources of the organizations in which digital curation occurs, the types of systems they have in place, the volume and types of information to be curated, and the degree to which curation tasks have been integrated into other workflows and business processes. These variations can be found not only between different sectors (e.g., financial, retail, entertainment, manufacturing, health care, research, and education), but also within organizations in the same sector.

Automation is already a common element of digital curation, and it has been increasing. As the field of digital curation matures and advances, the development and adoption of standards, norms, and best practices have enabled more automation of manual curation tasks. Investments in technology, software, and system enhancements—and in the people with the requisite knowledge and skills to design and build such systems—have also furthered automation.

How will automation affect future demand for a workforce in digital curation? Clearly, effective and long-term curation of digital information is a massive and complex undertaking. It could absorb an enormous amount of human capital. Manually mediated curation is labor intensive, sometimes prone to errors or bias, and therefore ultimately expensive. Automation of more aspects of digital curation may reduce dependence on manually mediated tasks, and thus reduce some of the increase in demand for human capital, in some areas of digital curation.

The creation of metadata is a prime candidate for automation. It is an immense and essential task. In some cases, the volume of metadata required for effective documentation greatly exceeds the volume of the data being described. Yet complete metadata are crucial for the analysis of data, as well as being a research resource itself. Despite its value, manual creation of comprehensive metadata has often been disdained by data producers, seen as a distraction from the main pursuit of research or analysis. Because of the volume of metadata needed, the costs of its manual creation, and both the feasibility and appeal of automated metadata creation, this is a very likely area for further automation in digital curation.

As digital curation becomes more systematized, with the emergence of standards, software, workflows, and other tools that reduce the need for human manual effort, other curatorial tasks are increasingly automated. Naming datasets, assigning identifiers, checking for and correcting errors, and maintaining redundant copies for backup and security are other digital curation tasks that have been and are increasingly automated (e.g., Lots of Copies Keep Stuff Safe [LOCKSS][3] and Digital Record Object Identification [DROID][4]).

Automating some digital curation activities has many advantages. Labor is by far the largest component of curation costs (see KRDS model in Section 2.8). Automation may also decrease curatorial errors and improve data quality. Automation may not only lower the cost and professional burden of metadata creation, but may even lessen the need for metadata. As artificial intelligence and other techniques get better at inferring or uncovering meaning, the required level of metadata tagging may decrease. It may be important as well to develop tools that implement standards at the time of data capture or creation, for example, electronic lab notebooks, and this might also affect the need for further curation.

---

[3] See http://www.lockss.org.
[4] See http://www.nationalarchives.gov.uk/information-management/projects-and-work/droid.htm.

More automation of metadata capture, unique identification of digital objects, error detection and correction, format migration, and the like will be necessary to contain the costs of curation. It is worth noting, however, that automation of digital curation within existing research or business processes and automation of repository functions will only go so far if the process of moving data from an active system to a repository, data center, or cloud service, is not addressed at the same time. Most efforts at automating digital curation to date have focused on either building automated digital curation workflows for business and research processes or for repository functions, but not both. As studies based on the KRDS model discussed above show, the largest expense for repositories are the processes of acquisition and ingestion.

The geospatial domain provides an example of some of the benefits and also the limitations of automation. In this domain, substantial efforts have been made over the past two decades to automate some of the functions of digital curation. Early geospatial pioneers spent large amounts of time in painstaking manual digitizing, but by the 1990s automated methods had become sufficiently sophisticated and reliable. Massive collections of paper documents, including maps and images, have been successfully digitized using automated methods. Today, manual digitizing techniques are no longer taught in most courses on geospatial technology. This curatorial task has been entirely automated.

Another task in digital curation in the geospatial domain, registration to the Earth's surface, has also been increasingly automated. The task arises frequently when images captured from satellites or aircraft must be registered accurately to allow their contents to be analyzed and combined with other data, thus greatly enhancing their value. To accomplish this task, a number of registration points are selected from the image and their Earth coordinates (often latitude and longitude) are entered into a geographic information system (GIS). A variety of techniques for rubber-sheeting, that is, stretching the image to match the Earth, are available. The task is especially difficult at global scales when registration points are not available, such as when a landscape lacks easily recognized features, as for example over the oceans or large areas of forest. Today, well-tested methods are widely available and capable of accurately warping something as crude as a photograph shot from a moving helicopter so that it matches existing maps or databases.

Other tasks of digital curation in the geospatial domain are less susceptible to automation. For example, it would be of great value to the community of scientists conducting research across many borders and languages if the systems of land classification used in different countries could be made interoperable. This is important in many domains, including research on global climate change, biological conservation, and agriculture. Toward that end, the INSPIRE (Infrastructure for Spatial Information in the European Community) project of the European Commission is attempting to harmonize the land classification systems of each member state. Simple cross-walks (Class A in Country 1 is the same as Class B in Country 2) are easy to do, but reality is often rather more complex (Class A in Country 1 is somewhat like Class B in Country 2). Standard methods and tools for this process, often termed semantic reasoning, are being adopted, but it is likely that the task will never be fully and satisfactorily automated.

Thus, some tasks of digital curation in the geospatial domain have been successfully automated; others will continue to require expert human judgment well into the future. Research has also produced a better understanding of the types of digital curation tasks that are not amenable to automation, such as those that require complex semantic reasoning, depend on inference to compensate for incomplete or inaccurate information, or anticipate novel ways to analyze or exploit digital resources. More broadly, the variability of progress toward automation in different aspects of digital curation in the geospatial domain suggests that progress beyond automation of generic curation tasks (metadata capture, integrity checks, error detection, etc.) may advance more effectively if handled on a domain-by-domain basis.

What, then, is the potential for automation and how will increased automation affect future demand for the workforce in digital curation? In the view of this committee, this will ultimately depend on how the work of digital curation is organized, which technical and organizational models for digital curation prevail, and what level of resources are invested in the development of the field. Clearly, the degree of human versus automated tasks in curation has important implications for the types of careers and jobs associated with digital curation and the education of the future workforce. If routine curation tasks can be automated, then it is possible that digital curation will require a smaller number of employees engaged in manual curation tasks. The experience in other areas where information technology is applied to traditional tasks, however, suggests that far from reducing employment, automation both allows specialists to advance to more sophisticated tasks and increases the overall level of employment. To achieve a high degree of automation in digital curation during the next decade, investments in standards and software development are essential. Those investments, in turn, are likely to generate demand for specialists who can design and build systems and applications that support digital curation, develop and implement standards, create and promulgate effective policies and best practices, and design and deliver the next generation of digital curation services.

3.6 Conclusions and Recommendations

Conclusion 3.1: Jobs involving digital curation exist along a continuum, from those for which almost all tasks focus on digital curation to those for which digital curation tasks arise occasionally in a job that is embedded in some other domain.

Conclusion 3.2: Although digital curation is not currently recognized by the Bureau of Labor Statistics in its Standard Occupational Classification, other sources of employment data identify the emergence and rapid rise of digital curation and associated job skills.

Conclusion 3.3: There is a paucity of data on the production of trained digital curation professionals and their career paths. Tracking employment openings, enrollments in professional education programs, and the placement and career trajectories of graduates from these programs would help balance supply with demand on a national scale.

Conclusion 3.4: The pace of automation and its potential impact on both the number and types of positions that require digital curation knowledge and skills is a great unknown. Automation of at least some digital curation tasks is desirable from a number of perspectives, and its potential has been demonstrated in several domains.

Conclusion 3.5: Enhanced educational opportunities and new curricula in digital curation can help to meet the rapidly growing demand. These opportunities can be developed at all levels and delivered through formal and informal educational processes. Digital learning materials that are accessible online, for example, may achieve broad exposure and possible rapid adoption of digital curation procedures.

Recommendation 3.1: Government agencies, private employers, and professional associations should develop better mechanisms to track the demand for individuals in jobs where digital curation is the primary focus. The Bureau of Labor Statistics should add a digital curation occupational title to the Standard Occupational Classification when it revises the SOC system in 2018. Recognition of digital curation as an occupational category would also help to strengthen the attention given to digital curation in workforce preparation.

Recommendation 3.2: Government agencies, private employers, and professional associations should also undertake a concerted effort to monitor the demand for digital curation knowledge and skills in positions that are primarily focused on other activities but include some curation tasks. The Office of Personnel Management should issue guidelines for specifying digital curation knowledge and skills that should be included in federal government position descriptions and job announcements. Private employers, professional associations, and scientific organizations should specify the digital curation knowledge and skills needed in positions that require them.

3.7 References

BLS (Bureau of Labor Statistics). Federal Register Notice. 2014. "Standard Occupational Classification (SOC)—Revision for 2018; Notice." Volume 79. Number 99 (May 22).

BLS. 2012a. *Bureau of Labor Statistics Standard Occupation Classification.* http://www.bls.gov/soc. Accessed August 16, 2012.

BLS. 2012b. Employment Projections 2010-2020. Last modified February 1, 2012. http://www.bls.gov/emp/.

Csorny, L. 2012. Computer occupational employment statistics and projections. Presented at public session of the Study Committee for Future Career Opportunities and Educational Requirements for Digital Curation, Board on Research Data and Information, National Research Council, Washington, DC, May 3.

Fox, G. 2012. Data analytics and its curricula. Presented at Microsoft *e*Science Workshop, Chicago, October 9.

Indeed.com. 2012. Job Trends. www.indeed.com/jobtrends. Accessed August 17, 2012.

Levanon, G., B. Colijn, B. Cheng, and M. Paterra. 2014. *From Not Enough Jobs to Not Enough Workers: What Retiring Baby Boomers and the Coming Labor Shortage Mean for Your Company.* Research Report R-1558-14-RR. The Conference Board, Inc., September.

Liddy, E. 2012. Digital curation as a core competency. Presented to the Symposium on Digital Curation in the Era of Big Data: Career Opportunities and Educational Requirements, Board on Research Data and Information, National Research Council, Washington, DC, July 19.

Lockard, C. B., and M. Wolf. 2012. Occupational employment projections to 2020. *Monthly Labor Review* (January):84-108.

Maatta, S. L. 2011. Explore the data. *Library Journal.* http://lj.libraryjournal.com/2011/10/placements-and-salaries/2011-survey/explore-the-data-2/.

Maatta, S L. 2012. A job by any other name| LJ's placements and salaries survey 2012. *Library Journal.* http://lj.libraryjournal.com/2012/10/placements-and-salaries/2012-survey/a-job-by-any-other-name-ljs-placements-salaries-survey-2012/.

Miller, S. 2012. The future of work. Presented to the Symposium on Digital Curation in the Era of Big Data: Career Opportunities and Educational Requirements, Board on Research Data and Information, National Research Council, Washington, DC, July 19.

Palmer, C. L., C. A. Thompson, K. S. Baker, and M. Senseney. 2014. Meeting data workforce needs: Indicators based on recent data curation placements. In iConference 2014 Proceedings, March 4-7, 2014, Berlin, Germany. http://hdl.handle.net/2142/47308.

Chapter 4

Preparing and Sustaining a Workforce for Digital Curation

Throughout this report, the organization of the work of digital curation has been discussed as a continuum. At the one end are a great variety of professions—research scientists, business analysts, even video editors and sound masters. Digital curation may be only an occasional task, but is nonetheless an essential part of their jobs. At the opposite end are specialists whose work consists primarily, if not exclusively, of the active management and enhancement of digital information assets for current and future use. That trope of a continuum will be continued here, in considering the education and training of a digital curation workforce. Preparing that workforce will require educational opportunities that span the entire continuum. This will include graduate-level education in digital curation for some, discrete study programs and certificates for others, perhaps supplementary courses inserted into established curricula in other fields, or exposure through online courses or conferences. Because the activities of digital curation are conducted along a continuum by a broad spectrum of workers, the educational opportunities appropriate for training the digital curation workforce also cover a large span.

As noted in Chapter 1, the two ends of the continuum may be very distant from each other, but they are also connected. This too has implications for preparing the workforce. Digital curation specialists will need some knowledge of the disciplines and domains in which the digital information they curate will be used. Without some familiarity with the problems to be addressed, the goals to be pursued, as well as the customary methods, nomenclature, and practices of the fields in which the digital information assets are used, curators will not be able to make good decisions as they manage and enhance those assets for current and future use. Similarly, those who conduct curatorial activities as only a small part of their work, will need some study and command of the knowledge and skills of digital curation, regardless of how well they are educated in their own domains.

This chapter is the committee's consideration of education across the entire continuum. It presents a vision of a proper educational program for a digital curation specialist. It also reflects on how a more limited program of study might be inserted or integrated into the preparation of other professions, whose practitioners will also have some responsibilities for digital curation. It then surveys the educational opportunities currently available along the entire continuum, for students training to be digital curation specialists, students acquiring some knowledge of curation while pursuing degrees in their other disciplines, and midcareer employees seeking to upgrade their skills. The chapter concludes by identifying next steps to be taken.

4.1 Envisioning the Education of Professional Digital Curators

The proper preparation of professional digital curators is receiving increased attention. One purpose of the National Research Council's "Symposium on Digital Curation in the Era of Big Data" (July 2012) was to explore views on educational requirements from a representative group of practicing experts and stakeholders from government, a range of scientific disciplines, the entertainment and computer industries, and research libraries. As part of that symposium, Elizabeth Liddy, Dean of the School of Information at Syracuse University, an iSchool with a strong track record in education for information professionals in e-science and data science, proposed core competencies for digital curation (Liddy, 2012).

The Library of Congress has also addressed the education of digital curators through its Digital Preservation Outreach and Education program, which aims to establish a national trainer network that will provide instruction for organizations seeking to preserve their digital content. One part of the program has been an effort to determine a baseline curriculum for digital preservation based on six concepts to identify, select, store, protect, manage, and provide.[1]

Perhaps the most in-depth effort to develop a graduate-level curriculum to prepare students to work in the field of digital curation was undertaken by the School of Information and Library Science at the University of North Carolina at Chapel Hill with funding from the Institute of Museum and Library Services. That effort resulted in the Digital Curation Curriculum (DigCCurr),[2] a comprehensive, structured curriculum organized along six dimensions:[3]

- Mandates, Values, and Principles;
- Functions and Skills;
- Professional, Disciplinary, Institutional, Organizational, or Cultural Context;
- Type of Resource;
- Prerequisite Knowledge; and
- Transition Point in Information Continuum.

The committee reviewed the output of the DigCCurr project, considered the workshops held during the course of this study, and also consulted other independent research. Informed by that material, the committee proposes the following 11 distinct knowledge and skill areas as essential to the education of professionals in the field of digital curation. The descriptions are not meant to be comprehensive, but rather, suggestive of the relevant range of expertise:

i. General background and abilities. An educational background that includes mathematics and science, enhanced by a disciplinary or domain specialization, provides a sound foundation for careers in digital curation. Given the multifaceted nature of digital curation, though, being able to cope with issues such as heterogeneity, complexity, and volume of data and information can be highly useful. Skills typically considered "soft" are likely to be important as well. These include the ability to communicate effectively, to work both independently and collaboratively, to question assumptions and innovate creative solutions, and to negotiate solutions involving competing priorities.

ii. Data practices. Rather than being appended to and distinct from the activities of an ongoing enterprise, digital curation needs to fit as seamlessly as possible within the organizational and institutional context of data production and use. Those providing digital curation services can benefit from understanding this context, including:

---

[1] See http://www.digitalpreservation.gov/education/curriculum.html.
[2] See http://ils.unc.edu/digccurr/index.html.
[3] See http://ils.unc.edu/digccurr/digccurr-matrix.html.

- Disciplinary, professional, and institutional practices;
- Research methods, instruments, tools, and protocols;
- Standards of evidence, quality, and uncertainty;
- Data types and formats for both quantitative and qualitative data;
- Data processing, transformation, and documentation processes; and
- Relevant standards for data, data models, metadata schemas, ontologies, and technologies.

iii. Data collection and management. This category involves proficiency in a wide array of diverse activities, starting with gathering and analyzing requirements, identifying and selecting data of interest, and developing effective processes for data acquisition or harvesting. Preparation of data for broader use would follow, including such activities as cleaning, normalizing, reformatting, and, perhaps, anonymizing information or related steps to preserve confidentiality. Additional complex processes are involved to prepare data for ingestion or deposit into a repository, including the generation of metadata aligned with relevant schemas and ontologies, the creation of unique identifiers for managing not only citations, but also versions, components, subsets, and other derivative products. Tracking provenance and documenting the appropriate contextual information to support long-term preservation (e.g., OAIS[4]) are also essential to curation aimed at reuse of data for new purposes. Support for data annotation and publication may also be required, as well as facilities for building and sustaining linkages to related literature and data and integration of functionality.[5] Processes for deselecting, removing, and destroying data that no longer satisfy retention criteria are likely to be a part of this process as well.

iv. Data analytics. Data analytics, or the ability to explore, extract, and validate new relationships or features from a body of quantitative or qualitative data, draws on repository resources and services (e.g., looking for undiscovered relationships among Linked Open Data). Digital curators can benefit from understanding how researchers conduct this work in order to facilitate it. Significant aspects of data analytics include:

- Research design,
- Sampling techniques,
- Hypothesis development and testing,
- Data mining,
- Information extraction,
- Statistics,
- Algorithmic thinking and programming, and
- Performance evaluation and risk analysis.

v. Presentation and visualization. While information presentation and visualization are typically viewed as the product, or output, of data analytics, they also have a role in the larger life cycle of digital repository services, to the extent that they can provide accessible insight into the nature of the curated resources. To that end, those providing digital curation services can

---

[4] Open Archival Information System, ISO 14721:2012, see
http://www.iso.org/iso/home/store/catalogue_ics/catalogue_detail_ics.htm?csnumber=57284.
[5] See, for example, Linked Open Data at http://linkeddata.org/.

benefit from an understanding of techniques used in presentation and visualization, including information design and contextualization, as well as the evaluation of products, algorithms, and specific programs.

vi. Archiving and preservation. The knowledge a digital curator ought to acquire regarding archiving and preservation overlaps in principle with that of a more traditional archivist, but includes a substantial number of areas of expertise that are not typically included in archival education. Technical areas include information technology skills such as managing a local server, a computer cluster, a cloud computing resource, or any combination of these. They also include a broad understanding of how archiving and preservation requirements are addressed for digital resources, which do not enjoy the permanence of physical artifacts and require, instead, sustained attention. Some of these skills include:

- Ensuring the integrity and security of digital resources;
- Authenticating those who seek access to the digital resources;
- Deploying appropriate approaches for long-term preservation, including media refreshment, emulation, migration, conversion, and canonicalization;
- Employing appropriate preservation models, such as OAIS, LOCKSS,[6] or PLANETS;[7]
- Assessing the trustworthiness of digital repositories using resources such as TRAC;[8] and
- Understanding the forensic role and responsibilities of digital repositories.

Furthermore, traditional archivists who typically are trained to work with historical sources for use by scholars in humanities and social sciences, genealogists, investigative journalists, and the like, will need deep immersion in the types of resources, methods, and data practices present in a much wider range of disciplines.

vii. Technologies, tools, and infrastructure. Digital curation involves building bridges from the producers of digital resources in many organizational contexts (e.g., in universities, government, industry, and the public) and who work in extraordinarily varied areas (e.g., research, finance, manufacturing, medicine, entertainment, and so on), to an equally varied population of current and future users. Those providing digital curation services confront a daunting array of technology-enabled choices, and need to understand and anticipate the implications of their decisions. A representative set of areas in which such knowledge can be identified includes:

- Data acquisition (e.g., from instrumentation, sensors, lab notebooks, and geographically enabled devices);
- Data modeling;
- Database design, construction, and management;
- Software development environments;
- Network architecture;
- Repository infrastructure;
- Web services;

---

[6] Lots of Copies Keep Stuff Safe; see http://www.lockss.org/.
[7] Preservation and Long-term Access through NETworked Services; see http://www.planets-project.eu/.
[8] Trusted Repository Audit Checklist; see http://www.crl.edu/archiving-preservation/digital-archives/metrics-assessing-and-certifying-0.

- Access systems;
- Preservation systems;
- Markup languages;
- System administration;
- Usability testing;
- Technology assessment; and
- Interoperability.

viii. Policy and planning. Digital curation also includes sufficient knowledge of relevant institutional, national, and international policies to ensure conformance with legal mandates, professional best practices, and expectations of practitioners at institutions or consortia. Among the more obvious of these are the protections afforded to intellectual property through copyright and digital rights management, which are often expressed with licensing or service agreements that restrict some access through terms of use, embargoes, or related provisions. Digital curation also inherits principles from the archives profession that may serve as a foundation for policies guiding collection, retention, preservation, recovery, and rescue. Furthermore, the management of any large-scale system involves expertise in risk assessment, disaster planning, and sustainability. The DRAMBORA Consortium[9] offers useful tools and guidance in assessing risk in digital repositories.

ix. Values and principles. Digital curation services ought to conform to the values and principles of the respective discipline, as well as those of the organization providing the services. The global expanse and ubiquity of network infrastructure, however, presupposes an underpinning of responsibility for principled activities grounded in fundamental values. The education of those working in digital curation is improved by the careful and deliberate attention to ethical, legal, cultural, and economic considerations that impact the production of knowledge and contribute to the evidentiary record. Digital curation providers will be accountable for the services offered, including conformance to relevant privacy provisions and the legitimate expectations of their users. The regulations, policies, norms, and values surrounding access, privacy, retention, repurposing, and manipulation of digital information are complex, ambiguous, and sometimes contradictory. Professionals engaged in digital curation should be prepared to analyze ethical dilemmas, identify conflicting principles and policies, and make informed recommendations or decisions to resolve such conflicts.

x. Services and support. As is true of many professions, service provision is at the heart of digital curation operations and key to their success, and covers a range of areas of responsibilities, including:

- Liaison and consulting;
- Instruction and training;
- Enhancement, including metadata, annotation, and linking;
- Information resource development;
- Outreach, advocacy, and promotion;
- Current awareness services (push and pull); and
- Support for virtual communities.

---

[9] Digital Repository Audit Method Based on Risk Assessment; see http://www.repositoryaudit.eu/.

xi. Management and administration. Likewise, those working in digital curation should be competent in basic management and administrative processes, including the following:

- Cost-benefit analysis,
- Strategic planning,
- Project management and planning,
- Staff development,
- Supervision,
- Training,
- Grant and report writing,
- Cross-institutional coordination,
- Expectation management and complaint handling.

4.2 Envisioning Education at the Other End of the Continuum

The program of study in digital curation envisioned above for training curatorial specialists would be neither feasible nor relevant for those aiming to conduct digital curation as but one part of their research or practice in other domains. What would be the best strategy for educating these other students? This is an important concern. For example, as many domain researchers producing digital data now learn about data management in an ad hoc way, they are not attending to preservation aspects of data management and are unaware of existing data services (Jahnke et al., 2012).

Providing students in disciplines generating and using digital content with the necessary digital curation knowledge and skills might be achieved by selectively integrating that content into their existing curricula. This approach, sometimes referred to as "microinsertion," would consist of introducing small amounts of new material on very specific curation topics into existing course materials, lectures, readings, exercises, and exams.[10] Materials for microinsertion might include lecture notes, reviews of the literature, examples of best practice, test questions, practical exercises, videos, or any other form of pedagogical material. Material would be modularized to ease its insertion into existing programs and could be made available via the Internet, with appropriate methods of search, evaluation, and retrieval.

For disciplines that are particularly data intensive, greater weight might be given to the study of digital curation. Beyond additional course material or even entire courses, some fields might conceive subspecialties or programs, perhaps resulting in a certificate of completion rather than a degree. Such programs would yield a level of proficiency in digital curation beyond what might be expected of a typical student in the field, though not at the level of a professional digital curator.

Existing curricula in science, business, medicine, engineering, and other fields are already crowded with required material. Adding further content in digital curation will require carefully developed strategies, demonstration that the value of this additional training exceeds whatever it might displace, and leadership within each domain to advocate for such changes.

---

[10] Microinsertion has also been advocated as a suitable way to include material on ethics in programs that have never in the past given sufficient attention to it (Davis, 2006).

4.3 Current Educational Opportunities for Students of Digital Curation

Many professional programs in library science, information science, archival science, and information management have responded to the immense growth of digital information and the need for its professional curation by creating programs in digital curation, digital preservation, digital libraries, and management of electronic resources. The most organized and concerted efforts to instill digital curation knowledge and skills are within professional schools of library and information science (LIS) or Information Schools (also known as iSchools). Many of the early efforts, identified in Gold's (2010) comprehensive time line of data curation initiatives in LIS schools have matured into established programs. A small but growing number of iSchools now offer concentrations or certificate programs in digital curation.

Education initiatives supported by the Institute of Museum and Library Services (IMLS) Laura Bush 21st Century Librarian Program have contributed significantly to building capacity in digital curation education. Below are summaries of some of the notable initiatives sponsored by this IMLS program:

- Beginning in 2006, the University of North Carolina at Chapel Hill (UNC) and the University of Illinois at Urbana-Champaign received multiple awards to build capacity in digital curation education. UNC DigCCurr initiatives in digital curation, discussed in Section 4.1, culminated in a new post-master's certificate in 2013. The University of Illinois' specialization in data curation began in 2007 and was extended from the sciences to the humanities in 2008. A second specialization in sociotechnical data analytics was launched in 2012.
- The University of Arizona received two awards to advance digital collections and curation education efforts in 2006 and 2009.
- In 2008, the University of Michigan's educational offerings were advanced by a collaboration to support internships and curriculum development in digital curation and preservation.
- Syracuse University started a program related to digital curation by developing training for e-science librarians in 2009, and introduced a certificate of advanced study in its data science program in 2012.
- The University of Tennessee received awards for training Ph.D.-level educators in science data and information in 2009, and for master's students in scientific data curation in 2011.
- In 2011, the University of North Texas began development of a post-master's graduate academic certificate in digital curation and data management.
- One of the newest efforts is the National Digital Stewardship Residency, a collaboration between IMLS and the Library of Congress Office of Strategic Initiatives to "build a dedicated community of professionals who will advance our nation's capabilities in managing, preserving, and making accessible the digital record of human achievement."[11]

The National Science Foundation (NSF) has also supported projects related to curation for many years and provided part of the early foundation for some of the efforts listed above. Of

---

[11] See http://www.digitalpreservation.gov/ndsr/.

particular note among NSF-supported initiatives is the ongoing Open Data Integrative Graduate Education and Research Traineeship (IGERT) at the University of Michigan.[12] The goals of the IGERT initiative are to promote conduct of responsible data-intensive science and engineering and to build a community of practice around open sharing and reuse of data in bioinformatics and materials science and engineering. The NSF has also funded other digital curation education initiatives. The new data science certificate program at Syracuse University was seeded by an NSF award for training cyberinfrastructure facilitators in 2008. The first step in the curriculum for the Specialization in Data Curation at Illinois was developed with an NSF award in 2006 for a Scientific Information Specialist program that evolved into a master's degree option in the campus bioinformatics program.

Despite these various initiatives, education options for digital information professionals remain limited even as demand for digital curation competencies is increasing in the job market. An analysis of LIS-oriented data curation programs identified only 16 institutions offering data curation courses (Harris-Pierce and Liu, 2012). The Data Curation Curriculum database,[13] which gathers information on programs and course descriptions related to data curation at LIS and iSchools, also suggests that educational opportunities are inadequate to the needs of training a skilled digital curation workforce (see Varvel et al., 2012, for details on objectives, methods, and limitations of the database[14]). In reviewing available information from the database on 475 courses in 158 separate programs at 53 institutions, the educational options appeared to be uneven, with limited opportunities for intensive digital curation preparation.

The limitations revealed by the Data Curation Curriculum database fall into several categories. A review of the course descriptions found in the database suggests that significant skill sets needed for digital curation, in areas such as metadata and digital preservation, are being covered in varying degrees. For example, of the approximately 60 courses covering metadata, about two-thirds explicitly considered digital information, with about one-third of those addressing digital research data. Only a few course descriptions specified coverage of scientific metadata standards, metadata for data management applications, or more general information modeling for digital content and data. Courses addressing preservation were even less aligned with a digital curation focus. Of the 62 course descriptions covering preservation, only 19 specified digital preservation as the primary subject of the course or a topic within the course. Project management course descriptions generally included planning and management for digital libraries or digital preservation, but only 14 such courses were identified across all schools.

Course descriptions contained in the Data Curation Curriculum database suggest that instruction in technology, statistics, and computer programming directly relevant to digital curation is also limited.[15] Other areas of importance to the field of digital curation but which were addressed by only a moderate number of courses include selection and appraisal, access

---

[12] See http://opendata.si.umich.edu/.

[13] See http://cirssweb.lis.illinois.edu/DCCourseScan1/.

[14] The methods are described at this website: http://dl.acm.org/citation.cfm?id=2132275.

[15] Course descriptions in the Data Curation Curriculum database reveal that among relevant technology courses, only 8 courses appeared to be exclusively devoted to database administration and design, and 13 systems analysis courses were identified. Statistics was explicitly covered in 16 courses, including topics on data mining, data analytics, descriptive and inferential statistics, and probability. Computer programming options were also limited. Typically in LIS programs, aspects of programming are embedded in a number of different kinds of courses, such as information processing and data mining. Students at most schools would have access to programming courses in computer science departments, but these classes would be unlikely to have an orientation to curation concerns, as in LIS courses that relate programming specifically to information resources and services.

and use, legal issues, and research methods. Further, the Data Curation Curriculum database furnishes no evidence of adequate adaptation of other relevant areas traditionally covered within LIS programs, such as information behavior, information retrieval, collection development, and information policy. Other more specific topics important for digital curation were rare and scattered within different kinds of courses.[16]

Overall, the educational opportunities for individuals wishing to pursue professional training as digital curators have grown. New programs, many supported by IMLS and NSF, have been established in recent years. Capacity in digital curation education is being built. Nonetheless, the more traditional training programs are only beginning to adapt their course offerings to the needs of digital curation professionals. Although many of the principles and skills covered in conventional degree programs are integral to digital curation education, courses continue to be too general in nature, with inadequate attention given to the specific knowledge and skills needed for curation of digital information.

4.4 Current Educational Opportunities for Students in Other Disciplines

Students pursuing courses of study in a wide range of disciplines increasingly need, and are being provided with, exposure to the knowledge and skills necessary for digital curation. In some cases this involves the microinsertion of additional course content, but it may include entire course modules or certificate programs that convey a more substantial grounding. At whatever level, students incorporate this work into their primary fields of study. In some instances, attention to both digital curation and a scientific discipline become entirely entwined. This is the case in several newly emerging hybrid fields.

Materials suitable for microinsertion are being developed in a number of disciplines. The Digital Library for Earth Science Education[17] was organized over a decade ago precisely to develop such materials. Often, minicourses in digital curation are created by academic libraries. These provide instruction in data management skills for domain students and researchers, as part of an acceleration of campus-based services for research data. Students are the primary audience at a number of institutions, with programs targeting undergraduate and graduate students in medical and health programs (Piorun et al., 2012) and engineering and science graduate students for instruction in "data information literacy" (Carlson et al., 2011).

One promising model for transitioning domain experts into curation work has been developed by the Council on Library and Information Resources/Digital Library Federation as an extension of their Postdoctoral Fellowship Program. Recent Ph.D.s from any natural or social science discipline are eligible for placements at host institutions in units engaged in curation operations and research. To date, most positions have been in research libraries or situated within collaborations among libraries and other university units, with a few opportunities in data centers.

Minicourses, microinsertion of material, and postdocs provide some training in digital curation to those studying other fields. Fuller attention to digital curation may be accomplished through more extensive programs offered within academic departments. Programs in Geographic Information Science (GIScience) provide an example. Such programs now exist in a large

---

[16] For example, only one course explicitly included data harvesting and aggregation. Two courses covered data quality, and isolated courses were identified on data manipulation and exploratory data analysis. The topic of data transformation was covered in two courses on "Organization of Information in Collections" and "Digital Library Implementation."

[17] See http://www.dlese.org.

number of institutions of higher education in the United States. Their purpose is to impart the skills and principles needed to work intensively with geographic (or geospatial) information, using technologies such as geographic information systems (GIS), satellite remote sensing, and geographically distributed sensor networks. Programs in GIScience have long included material related to the curation of geospatial data. It is common, for example, to find data sharing and metadata, the management of distributed databases, archiving, and online and cloud-based GIS, all discussed within the context of geospatial data. GIScience programs can be found within departments of geography, earth science, civil engineering, urban planning, and several other more traditional disciplines.

Increasingly, subspecialties addressing aspects of digital curation are evolving into certificate programs, either freestanding or housed within other academic departments. The committee reviewed 37 data-oriented programs compiled primarily from informal inventories (Fox, 2012; Varvel et al., 2011). These included programs embedded within computer science, informatics, business, and the sciences. The programs exist at the undergraduate, master's, and doctoral levels, with specializations including data analytics, business analytics, predictive analytics, data science, web science, and information technology and systems. Notable examples include:

- Certificate in Data Science at the University of Washington eScience Institute,
- Informatics degrees at Indiana University, and
- Information Technology and Web Science program at Rensselaer Polytechnic Institute.

Many data science programs emphasize statistics and analytics, with strong coverage of programming, databases, and data mining, with some attention devoted to areas such as system design and data management. Other prominent areas include information visualization and research methods. There appears to be limited coverage of storage, data processing and transformation, data collection, and ethics, all of which are important elements of competency in digital curation. Surprisingly, standards appeared to be strongly emphasized in only one program.

In some fields, students have the opportunity to integrate study in their chosen discipline with training in aspects of informatics. This is particularly the case at the graduate level, with many of the digital curation skills discussed in this report covered in graduate programs, as educational opportunities evolve in tandem with shifts toward data-intensive research and informatics. Research in certain fields, such as molecular biology, biodiversity, ecosystem studies, and climate science has become very data intensive, leading to hybrid specialties such as ecoinformatics, bioinformatics, biodiversity informatics, and climate informatics.

The emergence of the field of bioinformatics is a response to the need to manage the burgeoning data resources available in biology in order to take advantage of new analytical methods and techniques. Bioinformatics has grown rapidly as a career path, as evidenced by associated training programs. Currently, approximately 25 U.S. universities and colleges offer undergraduate programs in bioinformatics, most commonly as an area of concentration related to computer science, bioengineering, life sciences, or another field. At the graduate level, 19 universities offer doctoral programs and 32 offer master's programs with bioinformatics in the title. These include programs related to medical informatics, computational biology, and genetics. The content of these programs, displaying an integration of informatics and biology, is typified by the program of study required for a master's degree at The Johns Hopkins University:

- Core courses in molecular biology, genetics, algorithms, database concepts, and biological databases;
- Concentrations in informatics tools, protein structure data and proteomics, genome sequence data, microarray data, and semantic web studies; and
- Computer science courses including software engineering, XML (extensible markup language) design, data visualization, machine learning, distributed systems, and cloud computing.

Biodiversity informatics is another example of a hybrid field that has fully integrated the study of digital curation at all levels. For undergraduate students, the NSF has supported a Research Coordination Network in Undergraduate Biology Education called Advancing the Integration of Museums into Undergraduate Programs (AIM-UP!) to introduce museum-based informatics and data curation into the curriculum. Digital curation skills are increasingly taught as part of doctoral training in biodiversity research fields such as taxonomy and systematic biology. Short courses for established biodiversity researchers have also begun to appear over the past decade in some conferences in this and other fields.

Exposure to and practical experience with current digital curation practices can be valuable complements to formal programs of study. Unfortunately, practitioners often lack support or reward structures for participating in teaching and training. A few programs have developed promising strategies for providing students with internship opportunities (e.g., Kim et al., 2011). Other models for student field experiences include partnerships between academic programs and established national data centers to build on the advanced expertise developed within these institutions over decades (e.g., Kelly et al., 2013). Currently, such opportunities are available to a relatively small number of students.

4.5 Current Opportunities for Midcareer Employees

Attention is often accorded to the development of courses and curricula for students. What of the continued education and training of midcareer employees? These individuals also need opportunities to deepen or upgrade their skills and knowledge in digital curation.

A large portion of the current digital curation workforce consists of midcareer professionals, some of whom received their education when most information was still in analog form. What is certain is that data and computing were accelerating at a much slower pace, and print journals and books were considered the primary mechanisms for scholarly publication. Information professionals find that their training is rapidly dated, due to the rate of change in all aspects of information technology. In the scientific disciplines, many researchers with responsibilities for digital curation developed those skills without formal training, often following completion of graduate degrees.

Midcareer employees, practitioners, and researchers require opportunities to further their training in ways more flexible than traditional academic coursework and formal certification. Continuing education options for working professionals have been developed by a variety of institutions in various formats. These nondegree options include 1-day workshops, short courses, conferences, and longer institutionally based programs developed by iSchools and professional associations.

Two iSchools established regular institutes beginning in the mid-2000s, with DigCCurr at the University of North Carolina focusing on digital curation and the Summer Institute for Data

Curation at the University of Illinois focusing on curation of research data. These early efforts emphasized the full curation life cycle and best practices and tools for digital curation. More recently, the E-Science Institute, developed by the Association of Research Libraries, the Digital Library Federation, and DuraSpace, offered a series of learning modules to assist academic libraries in advancing an agenda for e-research support, with a particular focus on the sciences, cyberinfrastructure, and data curation. Another active technical community working in libraries, museums, and archives developed the CURATEcamp "unconference" series in 2010.

Efforts in the specific area of digital preservation have been sustained for over a decade. One initiative in this area is the Digital Preservation Management Workshops series.[18] This curriculum has three themes—organizational infrastructure, technological infrastructure, and requisite resources—for a target audience of managers of digital preservation programs in cultural institutions. The Library of Congress has also developed the Digital Preservation Outreach and Education program, as noted earlier in this report.

A number of professional organizations in the sciences are also active in fostering opportunities for continuing education for scientists in their disciplines. The Federation of Earth Science Information Partners (ESIP) and the American Geophysical Union (AGU) offer workshops and short courses on best practices in data management. The NSF-funded DataONE initiative has developed curriculum modules in data management. Emergent communities of practice that develop and share new knowledge also provide a kind of informal education through participation in their initiatives and the resources they generate. The Biodiversity Informatics Standards (TDWG) association (formerly, the Taxonomic Database Working Group), discussed in Chapter 2, is an example.

Employers are also concerned with upgrading the skill sets of their employees. Many seek to furnish in-house opportunities for professional development for their staff. Several programs are available for employers to use. One model program, developed by the Bibliothèque Nationale de France (Bermès and Fauduet, 2011), covers four areas: digital information, data models, project management, and long-term preservation. Topics such as formats, models, and standards for digital objects are highlighted, with an emphasis on areas that need to adapt to change, such as management, workflows, and proprietary rights.

Online continuing professional development is another attractive option for working professionals, and can range from webinars to executive degree programs. Among the more visible trends in higher education in recent years has been the introduction of Massive Open Online Courses (MOOCs). These are online courses offered free or at low cost by reputable institutions, and carry no course or degree credit. MOOCs may provide a suitable way to promote education in digital curation, particularly to small institutions with limited resources for professional development.

4.6 Looking Ahead

This chapter has provided the committee's vision of what will be necessary for the education of a workforce for digital curation, addressing the continuum from dedicated curatorial specialists to researchers and practitioners undertaking some curatorial activities. It has also surveyed current opportunities available to those pursuing education and training in digital

---

[18] Partially funded by grants from the National Endowment for the Humanities, the series was begun at Cornell University in 2003, moved to the Inter-university Consortium for Political and Social Research (ICPSR) at the University of Michigan in 2008, and is now hosted by the Massachusetts Institute of Technology.

curation along many different paths. Those opportunities will continue to evolve in response to the demand for training in digital curation.

Expertise requirements will clearly vary depending on application area, scale of operation, and work environment, with positions ranging from curation-centric roles in multidisciplinary repositories, archives, libraries, and data centers, to highly specialized discipline-, industry-, or application-specific roles (Hedstrom, 2012). Curation-centric positions will rely on the broad knowledge and skills underlying digital curation processes and technologies, with an emphasis on interoperability and the differing data requirements across disciplines. Specialized positions will demand more domain-specific knowledge and related informatics and computational expertise.

Of course, the full complement of knowledge and skills will rarely be found in one individual. An effective digital curation strategy is thus likely to engage the coordinated efforts of a team, leveraging the strengths of each member and balancing the best practices of digital curation with the intellectual traditions and requirements of the discipline. This will help ensure the appropriate accessibility and usability of the digital resources, for new communities of users in the future.

This sketch of the future is reasonable, yet not certain or complete. Different institutions, disciplines, domains, and sectors of the economy display very different levels of awareness of the need for and value of digital curation. Furthermore, the pace at which they will recognize the benefits of digital curation, adopt standards and best practices, and invest in both automated solutions and human personnel remains to be seen. Pressure for greater access to digital information in scientific research, education, industry, government, and cultural heritage suggests that there is an immediate need to build capacity in the nation's workforce to meet digital curation demands. This will not happen without resources and leadership.

During the next decade, there will be a particular need for leaders in a wide variety of organizations who can develop digital curation policies, programs, and technologies that reinforce each other and facilitate curation throughout the information life cycle. Those leaders will need to rely on professional curators who will create standards and best practices, monitor developments in the field, solve problems that result from new technologies or disruptive technological change, and train others in digital curation.

4.7 Building on Current Foundations

The importance of digital curation has long been recognized by some. As we noted already in Chapter 2, a substantial body of reports and studies have focused on digital curation (e.g., Lord and Macdonald, 2003; Swan and Brown, 2008; Interagency Working Group on Digital Data, 2009; National Science Board, 2005; Blue Ribbon Task Force on Sustainable Digital Preservation and Access, 2010; Auckland, 2012; Lyon, 2012). As the accumulation of digital information continues to increase exponentially, becoming both ubiquitous throughout society and critical to its functioning, the necessity for rigorous curatorial processes will become profoundly apparent. It is important for those in a position to influence this eventuality to lead the coordinated effort at resource mobilization, technological innovation, and education to prepare the workforce for effective, scalable, and affordable digital curation. Future efforts will build on the well-established foundations and service orientations of research libraries and archives, data centers, government agencies, research communities and commercial service providers.

As presented above, early initiatives sponsored by the federal government and conducted by LIS programs and iSchools produced courses, concentrations within degree programs, and continuing education for practitioners. More recently, awareness of the need for curation services for digital research data has increased due to further federal actions, including the introduction of guidelines from funding agencies requiring research proposals to include data management plans, and the February 2013 memorandum issued by the Executive Office of the President, Office of Science and Technology Policy (OSTP)—*Increasing Access to the Results of Federally Funded Scientific Research* (Holdren, 2013), which covers both peer-reviewed publications and digital data. Progress in digital curation has benefited from the very substantial attention of NSF (from the disciplinary perspective of the sciences) and IMLS (from the professional perspective of library and information science). Sustaining this progress during a period of rapid change and growth is essential.

4.8 Many Stakeholders of Progress

Beyond NSF and IMLS, many other federal agencies and departments have a fundamental responsibility to carry out well-informed digital curation, and therefore have a stake in the development of a strong, expert workforce in digital curation. They include the Library of Congress, the National Archives and Records Administration, the National Institutes of Health, the Department of Defense, the National Aeronautics and Space Administration, the Department of Energy, and the Census Bureau, to name only a few. Indeed, all government agencies can be encouraged to participate in the advancement of digital curation from their mission-oriented perspectives and to integrate the evolving state of the art into institutional policies and practices. The OSTP may be able to provide coordinating leadership across the federal government.

The private sector also has much to gain from effective and consistent digital curation processes, policies, and procedures, and therefore in the preparation of the digital curation workforce. Many businesses have major investments in digital information assets; some industries are entirely dependent on them. Business schools, which increasingly include data analytics and data mining in their curricula, also have a responsibility to educate business leaders about the value and need for digital curation. Moreover, an inevitable result of advances in digital curation will include proposed modifications to existing standards and proposals for new standards. Expertise in digital curation within the relevant standards bodies and professional associations will help ensure that the interests of all parties, private and public, are addressed.

4.9 Next Steps

Educating and training a workforce for digital curation will take resources and leadership. A sustained and targeted campaign could engage the full spectrum of relevant stakeholder communities. Possible sponsors of such a campaign would have to truly understand the extent of the challenge to speak with authority and conviction. The national libraries (particularly the Library of Congress and the National Library of Medicine) are natural candidates. They have the reputation and a record of achievement in related areas. The momentum of the campaign could be increased by enlisting other organizations. For example, relevant professional associations (e.g., Association of Research Libraries,[19] American Library Association,[20] Special Libraries

---

[19] See http://www.arl.org.
[20] See http://www.ala.org.

Association,[21] Coalition for Networked Information,[22] Association for Information Science and Technology,[23] American Association for the Advancement of Science,[24] and EDUCAUSE[25]) can make such outreach part of their respective missions. The iSchools can build awareness among faculty, students, and alumni working in the field. The campaign would benefit from a bold and memorable tagline, perhaps something along the lines of "Dire Digital Stakes: Past, Present, and Future at Risk."

In addition to a bold campaign, incremental steps will also contribute to progress toward a well-prepared workforce in digital curation. One such step is the recruitment of more students with backgrounds in the sciences into graduate programs for digital curation specialists. For the most part, students currently applying to graduate programs in library and information science earned their undergraduate degrees in the humanities and social sciences. A survey of 3,507 recent graduates of 39 LIS programs in North America (2000-2009) indicates that only 25 percent of the students had Bachelor of Science degrees (Marshall, 2012). An earlier survey of LIS graduates in North Carolina indicated that 242 of 2,633 students had science backgrounds, with 80 in applied science, about 60 each in mathematics and life sciences, and about 20 each in earth and physical sciences.

The rapid transition to data-driven science and business analytics would benefit from a curatorial workforce that is knowledgeable and proficient in the sciences as well as in digital curation. Educators in existing data curation and data science programs have reported difficulties recruiting students, especially those with a background in the domain sciences. Attracting students with science backgrounds into iSchools will likely require broader awareness of the intrinsic challenges and rewards of digital curation, its essential role in advancing scientific research, and the career options in the field (Weber et al., 2012).

To recruit and retain a quality workforce, a career path needs to be attractive and visible (Gregg, 2012). Career advancement might involve a path that crosses a range of professional contexts, including "small data, big data, individuals, small research teams, large corporate endeavors" and across organizations (Thomas, 2012). Professional opportunities will increase if skilled workers are able to move both within and across sectors.

The growth of communities of practice in the field of digital curation will also be essential for sustaining career opportunities and development. A professional community is forming, fostered in part by conferences and workshops such as the International Digital Curation Conference, the Committee on Data for Science and Technology—World Data System conferences, the Research Data Alliance Plenary meetings, the ASIS&T Research Data Access and Preservation Summit, and iPRES. Recurring institute training programs and sessions devoted to data science and informatics at many disciplinary conferences also help to build the professional community. A coordinating body would be useful to guard against fragmentation of this community, to support cross-fertilization across the growing education initiatives, and to ensure that education programs address emerging needs in research and practice.[26]

---

[21] See http://www.sla.org.
[22] See http://www.cni.org.
[23] See https://www.asis.org.
[24] See http://www.aaas.org.
[25] See http://www.educause.edu.
[26] An excellent model for a discipline-specific coordinating body for an increasingly active and distributed digital curation community is the International Society for Biocuration, a not-for-profit organization for biocurators, developers, and researchers that promotes the biocuration field and serves as a forum for the exchange of information (Howe, et. al. 2008).

The digital curation workforce also includes a role for citizen science. The increase of citizen science has contributed substantially to various types of research, from collecting data (e.g., the Audubon Bird Count)[27] to annotating them (cf. Flickr)[28] and processing them (cf. SETI@home).[29] To involve the public, the challenge is to identify tractable opportunities in digital curation and to make them accessible to the relevant communities of interest on a distributed basis. Public libraries can fulfill an important role in outreach, education, and fostering contributions from the public, such as metadata or tagging by amateur scientists, history enthusiasts, and others who have curatorial expertise.

One further step to be taken is the articulation of a research agenda. Digital curation continues to advance. It does so within a continually changing context. Thus the preparation of its workforce will also be in flux. Research is needed to investigate what challenges exist, how best to meet them, and how to translate those strategies into standard practice. Input to that research agenda should come from the full continuum of those engaged in digital curation activities, from curatorial specialists through to domain experts and industry practitioners.

4.10 Conclusions and Recommendations

Conclusion 4.1: Although the number and breadth of educational opportunities supporting digital curation have grown, existing capacity is low, especially for the initial education of professional digital curators and the midcareer training of professionals in other fields. In particular:

- Graduate and postgraduate certificate programs for educating professional digital curators (e.g., in LIS schools and iSchools) are expanding, but workforce demand is projected to exceed the output of existing programs.
- Midcareer practitioners with little or no formal education in digital curation rely on a spectrum of types of training, including online and in-person, experimental and time-tested, and just-in-time training, but this too is not sufficiently developed.

Conclusion 4.2: The knowledge and skills required of those engaged in digital curation are dynamic and highly interdisciplinary. They include an integrated understanding of computing and information science, librarianship, archival practice, and the disciplines and domains generating and using data. Additional knowledge and skills for effective digital curation are emerging in response to data-driven scholarship. More specifically:

- Individuals with an undergraduate degree in science, technology, engineering, or mathematics (STEM) disciplines and graduate-level education in digital curation are—and will continue to be—in particular demand as digital curators.
- Discipline specialists with informatics and digital curation expertise are, and will continue to be, in demand to provide discipline-focused curation services.
- Although the multidisciplinary character of digital curation as a career currently suggests a graduate education level, some knowledge and skills may be acquired through 2-year associate or 4-year bachelor's degrees.

---

[27] See http://birds.audubon.org/great-backyard-bird-count.
[28] See http://www.flickr.com.
[29] See http://setiathome.berkeley.edu.

- Continuing professional education alternatives will need to be flexible and diverse, providing a range of introductory and more specialized options through several modes of delivery, such as workshops, tutorials, online course modules, and webinars.

Conclusion 4.3: The range of needs and opportunities in digital curation, particularly when reflected in Office of Personnel Management position descriptions and Bureau of Labor Statistics descriptions of occupations, will require building and advancing a diverse community supported by a core of professionals and practitioners.

Recommendation 4.1: OSTP should convene relevant federal organizations, professional associations, and private foundations to encourage the development of model curricula, training programs and instructional materials, and career paths that advance digital curation as a recognized academic and professional discipline.

Recommendation 4.2: Educators in institutions offering professional education in digital curation should create cross-domain partnerships with educators, scholars, and practitioners in data-intensive disciplines and established data centers. The goals of these partnerships would be to accelerate the definition of best practices and guiding principles as they evolve and mature, to help ensure that educational and training opportunities meet the needs of scientists in specific disciplines, analysts in different business sectors, and members of other communities utilizing digital curation systems and services.

Recommendation 4.3: Federal agencies, private foundations, and industrial research organizations should foster research on digital curation that makes fundamental progress on problems with practical applications in their respective domains. Initial activities should focus on establishing research priorities and baseline analyses, including engagement and outreach through

- Conferences and symposia designed to recognize and communicate the need for, benefits of, and successes in digital curation; and
- Workshops for researchers in the public and private sectors to develop coordinated research agendas focused on enhancing the value and utility of digital resources, including metadata, interoperability, and automation.

The resulting agendas for research in digital curation should be tightly coupled with the curricula and offerings of educational programs to shape the field during a time of dynamic and dramatic growth and change.

4.11 References

Auckland, M. 2012. *Re-skilling for Research: An Investigation into the Role and Skills of Subject and Liaison Librarians Required to Effectively Support the Evolving Information Needs of Researchers.* Research Libraries UK. http://www.rluk.ac.uk/files/RLUK%20Re-skilling.pdf.

Bermès, E., and L. Fauduet. 2011. The human face of digital preservation: Organizational and staff challenges, and initiatives at the Bibliothèque Nationale de France. *International Journal of Digital Curation* 6(1):226-237.

Blue Ribbon Task Force on Sustainable Digital Preservation and Access. 2010. *Sustainable Economics for a Digital Planet: Ensuring Long-Term Access to Digital Information.* http://brtf.sdsc.edu/.

Carlson, J. R., M. Fosmire, C. Miller, and M. Sapp Nelson. 2011. Determining data information literacy needs: A study of students and research faculty. *Portal: Libraries and the Academy* 11(2):629-657.

Davis, M. 2006. Integrating ethics into technical courses: Micro-insertion. *Science and Engineering Ethics* 12(4):717-730.

Fox, G. 2012. Data analytics and its curricula. Presented at Microsoft eScience Workshop, Chicago, October 9.

Gold, A. 2010. Data curation and libraries: Short-term developments, long-term prospects. *Office of the Dean (Library)* 27. http://works.bepress.com/agold01/9.

Gregg, M. 2012. Workforce demand and career opportunities: Scientific data centers. Presented to the Symposium on Digital Curation in the Era of Big Data: Career Opportunities and Educational Requirements, National Research Council, Washington, DC, July 19.

Harris-Pierce, R. L., and Y. Q. Liu. 2012. Is data curation education at library and information science schools in North America adequate? *New Library World* 113(11 and 12):598-613.

Hedstrom, M. 2012. Digital data curation—workforce demand and educational needs for digital data curators. In *Trusted Digital Repositories & Trusted Professionals International Conference Proceedings*, Florence, Italy.

Holdren, J. P. 2013. Increasing Access to the Results of Federally Funded Scientific Research. Memorandum for the Heads of Executive Departments and Agencies. Office of Science and Technology Policy. http://www.whitehouse.gov/sites/default/files/microsites/ostp/ostp_public_access_memo_2013.pdf.

Howe, D., M. Costanzo, P. Fey, T. Gojobori, L. Hannick, W. Hide, D. P. Hill, R. Kania, M. Schaeffer, S. St Pierre, S. Twigger, O. White, and S. Y. Rhee. 2008. Big data: The future of biocuration. *Nature* 455(7209):47-50.

Jahnke, L., A. Asher, and S. D. C. Keralis. 2012. *The Problem of Data.* CLIR Publication 154. Council on Library and Information Resources, Washington, DC. http://www.clir.org/pubs/reports/pub154.

Kelly, K., C. L. Palmer, V. E. Varvel, Jr., S. Allard, C. Tenopir, M. S. Mayernik, and M. Marlino. 2013. Model development for scientific data curation education. In *Proceedings of the 8th International Digital Curation Conference,* Amsterdam.

Kim, Y., B. K. Addom, and J. M. Stanton. 2011. Education for eScience professionals: Integrating data curation and cyberinfrastructure. *International Journal of Digital Curation* 6(1):125-138.

Liddy, E. 2012. Digital curation as a core competency. Presented to the Symposium on Digital Curation in the Era of Big Data: Career Opportunities and Educational Requirements, Board on Research Data and Information, National Research Council, Washington, DC, July 19.

Lord, P., and A. Macdonald. 2003. *e-Science Curation Report: Data Curation for e-Science in the UK: An Audit to Establish Requirements for Future Curation and Provision.* Digital Archiving Consultancy Limited. http://www.jisc.ac.uk/uploaded_documents/e-ScienceReportFinal.pdf.

Lyon, L. 2012. The informatics transform: Re-engineering libraries for the data decade. *International Journal of Digital Curation* 7(1):126-138. http://ijdc.net/index.php/ijdc/article/view/210/279.

Marshall, J. G. 2012. *WILIS 2*, Ver. 9. University of North Carolina Odum Institute for Research in Social Science.

National Science Board. 2005. *Long-Lived Digital Data Collections: Enabling Research and Education in the 21st Century.* Washington, DC: National Science Foundation. http://www.nsf.gov/pubs/2005/nsb0540/.

Piorun, M. E., D. Kafel, T. Leger-Hornby, S. Najafi, E. R. Martin, P. Colombo, and N. R. LaPelle. 2012. Teaching research data management: An undergraduate/graduate curriculum. *Journal of eScience Librarianship* 1(1):8.

Rappa, M. 2012. Education for data scientists. Presented to the Symposium on Digital Curation in the Era of Big Data: Career Opportunities and Educational Requirements, National Research Council, Washington, DC, July 19.

Raskino, M. 2013. 5 Facts About Chief Data Officers. http://blogs.gartner.com/mark_raskino/2013/11/06/5-facts-about-chief-data-officers/Swan, A., and S. Brown. 2008. *Skills, Role & Career Structure of Data Scientists and Curators: An Assessment of Current Practice and Future Needs.* Bristol, UK: JISC http://www.jisc.ac.uk/publications/reports/2008/dataskillscareersfinalreport.aspx.

Thomas, C. 2012. Views of the sponsors: Institute of Museum and Library Services Scientific Data Centers. Presented to the Symposium on Digital Curation in the Era of Big Data: Career Opportunities and Educational Requirements, National Research Council, Washington, DC, July 19.

Varvel, V. E. Jr., C. L. Palmer, T. C. Chao, and S. Sacchi. 2011. *Report from the Research Data Workforce Summit, December 6, 2010, Chicago, IL.* https://www.ideals.illinois.edu/bitstream/handle/2142/25830/RDWS_Report_Final.pdf.

Varvel, V. E. Jr., E. J. Bammerlin, and C. L. Palmer. 2012. Education for data professionals: A study of current courses and programs. Pp. 527-529 in *Proceedings of the 2012 iConference*. New York: ACM.

Weber, N. M., C L. Palmer, and T C. Chao. 2012. Current trends and future directions in data curation research and education. *Journal of Web Librarianship* 6(4):305-320.

APPENDIX A

*Symposium on Digital Curation in the Era of Big Data:
Career Opportunities and Educational Requirements*

**Board on Research Data and Information
National Research Council
Keck 100
500 Fifth Street, NW
Washington, D.C.**

Thursday, July 19, 2012

AGENDA

Session 1—Why Is Digital Curation Important for Workforce and Economic Development?

Session Chair: Margaret Hedstrom, Associate Dean for Academic Programs, School of Information, University of Michigan

8:30   Chair's opening remarks—Margaret Hedstrom

8:35   Views of the sponsors
- Institute of Museum and Library Services—Susan Hildreth, Director
- National Science Foundation—Alan Blatecky, Director, Office of Cyberinfrastructure
- Alfred P. Sloan Foundation—Joshua Greenberg, Program Director

9:05   Keynote: Digital Curation and Big Data—David Weinberger, Senior Researcher, Berkman Center for Internet & Society, Harvard University

9:35   A Government Policy Perspective—Michael Stebbins, Assistant Director for Biotechnology, White House Office of Science and Technology Policy

9:50   National Science Foundation Policies About Information Access—Myron P. Gutmann, Assistant Director, Directorate for the Social, Behavioral, and Economic Sciences

10:05   *Coffee break*

Session 2—*Workforce Demand and Career Opportunities*
Session Chair: Michael Goodchild, Professor, University of California, Santa Barbara

10:30   Universities and Research Libraries—Anne Kenney, Director, Cornell University Library

10:45   Scientific Data Centers—Margarita Gregg, Director, National Oceanographic Data Center, National Oceanographic and Atmospheric Administration

11:00   A Data Scientist Perspective—Vicki Ferrini, Associate Research Scientist, Lamont-Doherty Earth Observatory, Columbia University

11:15   Entertainment Industry—Andy Maltz, Director, Science and Technology, Academy of Motion Picture Arts and Sciences

11:30   Panel Discussion

12:00   *Lunch*

Session 3—*Education and Training—Needs and Opportunities*
Session Chair: Carole Palmer, Director of the Center for Informatics Research in Science and Scholarship, University of Illinois at Urbana-Champaign

1:15   Digital Curation as a Core Competency—Elizabeth Liddy, Dean, School of Information, Syracuse University

1:30   Continuing Education—Nancy McGovern, Head, Curation and Preservation Services, MIT Library

1:45   Retooling the Existing Workforce—Steven Miller, Program Director, Skills & Community Information Management, IBM

2:00   Lessons Learned: the Case of Bioinformatics—Lawrence Hunter, Director, Center for Computational Pharmacology and Director, Computational Bioscience Program, University of Colorado, Denver

2:15   Education for Data Scientists—Michael Rappa, Founding Director, Institute of Advanced Analytics, North Carolina State University

2:30   Panel discussion

3:00   *Coffee Break*

Session 4—Paths Forward
Session Chair: Lee Dirks, Director of Education and Scholarly Communication, Microsoft Research

3:30   Summary of Workforce and Education Issues (Chairs of Sessions 1, 2, and 3)

4:00   Plenary Discussion

4:45   Symposium Chair's closing remarks—Margaret Hedstrom

5:00   *End of meeting*

APPENDIX B

STUDY COMMITTEE FOR FUTURE CAREER OPPORTUNITIES AND EDUCATIONAL REQUIREMENTS FOR DIGITAL CURATION

Margaret Hedstrom (Chair)
University of Michigan

Dr. Margaret Hedstrom is the Robert M. Warner Collegiate Professor in the School of Information at the University of Michigan (UM). Before joining the UM faculty in 1995, she was chief of state records advisory services and director of the Center for Electronic Records at the New York State Archives and Records Administration (1985-1995). Her current research interests include scientific data curation, archiving, and reuse. She was a member of the National Academy of Sciences Board on Research Data and Information. She has also served on the following NAS Committees: Committee to Study Digital Archiving and the National Archives and Records Administration (Member, 6/1/2002–6/30/2005) and Committee on Information Technology Strategy for the Library of Congress (Member; 1/18/1999–6/30/2001). She holds a Ph.D. in history and a M.L.S. in library science and information science, both from the University of Wisconsin-Madison. She is a Fellow of the Society of American Archivists.

Peter Fox
Rensselaer Polytechnic Institute

Dr. Peter Fox is Tetherless World Research Constellation Chair and Professor in the Earth and Environmental and Computer Sciences Departments at Rensselaer Polytechnic Institute. He joined the Tetherless World Constellation in 2008. Formerly, he was the chief computational scientist at the High Altitude Observatory of the National Center for Atmospheric Research. Professor Fox's research covers the fields of ocean and environmental informatics, computational and computer science, semantic data frameworks, and solar and solar-terrestrial physics. He is the past chair of the American Geophysical Union Special Focus Group on Earth and Space Science Informatics. He also was a member of the ad-hoc International Council for Science's Strategic Coordinating Committee for Information and Data and is the chair of the International Union of Geodesy and Geophysics's Union Commission on Data and Information.

He currently serves on the Board of Directors for the not-for-profit Open Source Project for a Network Data Access Protocol (OPeNDAP). Professor Fox holds a B.S. in mathematics (Honors) and a Ph.D. in applied mathematics from Monash University, Australia. He also held a postdoctoral position in astronomy at Yale University.

Michael F. Goodchild
University of California, Santa Barbara

Dr. Michael Goodchild (NAS) is professor emeritus of geography at the University of California, Santa Barbara (UCSB) and director of UCSB's Center for Spatial Studies. His research interests include urban and economic geography, geographic information systems, and spatial analysis. He was elected as a member of the National Academy of Sciences and as a foreign member of the Royal Society of Canada (2002), member of the American Academy of Arts and Sciences (2006), and foreign member of the Royal Society and corresponding fellow of the British Academy (2010). In 2007, he received the Prix Vautrin Lud. He has published more than 15 books and 400 articles. He serves on various NAS committees, including the National Research Council Board on Research Data and Information. He chairs the Advisory Committee on Social, Behavioral, and Economic Sciences of the National Science Foundation. He received his B.A. in physics from Cambridge University in 1965 and his Ph.D. in geography from McMaster University in 1969, and has received four honorary doctorates.

Heather Joseph
Scholarly Publishing and Academic Resources Coalition

Ms. Heather Joseph has served as the executive director of the Scholarly Publishing and Academic Resources Coalition (SPARC) since 2005. In that capacity, she works to support the broadening of access to the results of scholarly research through enabling open-access publishing, archiving, and policies on local, national, and international levels. Prior to her tenure at SPARC, she spent 15 years as a publisher in both commercial and not-for-profit publishing organizations. She was the publishing director at the American Society for Cell Biology, which became the first journal to commit its full content to the National Institutes of Health's pioneering open repository, PubMed Central, and she subsequently served on the National Advisory Committee for the project. She currently serves on the Board of Directors of the Public Library of Science, DuraSpace, and Impact Story. Ms. Joseph has served on the NAS Planning Committee for the Workshop on Strategies for Open Access and Preservation of Digital Scientific Data Resources in China: Challenges and Opportunities (5/21/2004–12/31/2004). Ms. Joseph recently completed a term as the elected president of the Society for Scholarly Publishing. She holds a B.S. in journalism and an M.A. in business administration, both from the University of Maryland.

Ronald L. Larsen
University of Pittsburgh

Dr. Ronald L. Larsen is a professor and dean at the School of Information Sciences (SIS) at the University of Pittsburgh. He has led a number of studies for the National Science Foundation, helping to develop research priorities in digital libraries and information management. During the mid to late 1990s, he was the assistant director of the Information Technology Office at the Defense Advanced Research Projects Agency (DARPA), where he led research programs in digital libraries, information management, and cross-lingual information utilization, with particular emphases on interoperability and the development of performance metrics for large-scale distributed information systems. His career includes 17 years at the University of Maryland, where he served as assistant vice president for computing, associate director of libraries for information technology, executive director of a 10-university consortium on workforce development, and affiliate associate professor of computer science. Dr. Larsen holds a B.S in engineering sciences from Purdue University, an M.S. in applied physics from Catholic University of America, and a Ph.D. in computer science from the University of Maryland, College Park.

Carole L. Palmer
University of Washington

Dr. Carole Palmer is professor in the Information School at the University of Washington. Her research investigates how to optimize cross-disciplinary access and the reuse value of data resources. As an educator, she has been a leader in data curation workforce development for nearly a decade, recognized in 2013 with the Information Science Teacher of the Year Award from the Association for Information Science & Technology. She leads collaborative teams of information scientists, domain researchers, and data archiving experts to address fundamental problems in building shared data resources and evolve best practices in data services. She is currently the principal investigator (PI) on the Site Based Data Curation project and the Data Curation Education in Research Centers project. She was co-PI on the Data Conservancy initiative from 2009 to 2012. She served on the NAS Committee on Building Cyberinfrastructure for Combustion Research (2009-2010). She holds an M.L.S. from Vanderbilt University and a Ph.D. in library and information science from the University of Illinois at Urbana-Champaign. She was the Director of the Center for Informatics Research in Science & Scholarship in the Graduate School of Library and Information Science at the University of Illinois from 2007 to 2014, in addition to her faculty appointment from 1996 to 2014. At the University of Washington she is leading development of the data curation initiatives in the Information School and is an affiliate of the eScience Institute.

Steven Ruggles
University of Minnesota, Minneapolis

Dr. Steven Ruggles is a professor of history and population studies at the University of Minnesota, and the director of the Minnesota Population Center. His research interests are in historical demography and large-scale population data infrastructure, especially methods for data integration and dissemination. He is best known as the creator of the Integrated Public Use Microdata Series (IPUMS), the world's largest population database, and has made important contributions to the study of long-run demographic changes, focusing especially on changes in the family. Dr. Ruggles received the William J. Goode Award from the Family Section of the American Sociological Association, the Allen Sharlin Award from the Social Science History

Association, the Robert J. Lapham Award from the Population Association of America, and the Warren E. Miller Award from the Inter-university Consortium for Political and Social Research. He currently serves as a member of the Census Scientific Advisory Committee and the National Science Foundation Advisory Committee for the Social, Behavioral, and Economic Sciences, and he is the 2015 president of the Population Association of America. He also has written extensively on demographic change and population data. Dr. Ruggles earned his B.A. in history from the University of Wisconsin in 1978, and his M.A. and Ph.D. from the University of Pennsylvania in 1984, whereupon he did postdoctoral training in sociology and demography at the University of Wisconsin.

David E. Schindel
Smithsonian Institution

Dr. David Schindel is the executive secretary of the Consortium for the Barcode of Life, an international initiative hosted by the Smithsonian Institution's Museum of Natural History. He was a member of Yale University's Department of Geology and Geophysics and was curator of invertebrate fossils in the Yale Peabody Museum from 1978 to 1986. In 1986, Dr. Schindel joined the National Science Foundation where he directed a variety of funding programs that provided support for research in systematic biology and for improving facilities and constructing specimen databases in natural history museums and herbaria. In 1997, Dr. Schindel worked in the U.S. Senate as a Brookings Institution LEGIS Fellow in the office of Senator Jeff Bingaman. From 1998 to 2004, Dr. Schindel served as the National Science Foundation's European representative. He is currently active on the following NAS committees: U.S. Delegation to the Pacific Science Association Council, June 2011 (Chair, 5/20/2011–12/31/2011) and U.S. National Committee for the Pacific Science Association (Member, 1/1/2010–12/31/2011). Dr. Schindel was trained as an invertebrate paleontologist and holds a Ph.D. in geological sciences from Harvard University.

Stephen Wandner
The Urban Institute

Dr. Stephen Wandner is a visiting scholar at the Urban Institute in Washington, D.C. He is also a visiting scholar at the W. E. Upjohn Institute for Employment Research. He worked for the U.S. Department of Labor for many years. At the Unemployment Insurance Service in the Employment and Training Administration of the U.S. Department of Labor, he was an actuary, director of benefit financing, and deputy director of the Office of Legislation, Research and Actuarial Services. For the Employment and Training Administration, he was the director of Research and Demonstrations and director of Strategic Planning. Dr. Wandner has published over 30 articles and 4 books. Princeton University's Industrial Relations Section awarded him the Richard A. Lester Prize for the Outstanding Book in Labor Economics and Industrial Relations, published in 2010, for *Solving the Reemployment Puzzle: From Research to Policy*. He was awarded the 2011 Outstanding Practitioner Award in recognition of outstanding contributions to research and practice in the field of employment relations by the Labor and Employment Relations Association. His recent research includes development of education and training scorecards using alternative employment and earnings data sources; an analysis for AARP of pubic workforce programs for older workers; an evaluation of the Transition

Assistance Program employment workshop for military service members; an implementation analysis of workforce programs funded by the American Recovery and Reinvestment Act; an analysis of the response of the public workforce system to the Great Recession, funded by the Rockefeller Foundation; and a short-time compensation employer survey that is part of an evaluation of that program. He received his Ph.D. in economics from Indiana University.